Food Process Monitoring Systems

Food Process Monitoring Systems

Edited by

A.C. PINDER
Head of Biophysics Group
AFRC Institute of Food Research
Norwich

and

G. GODFREY
Process Development Manager
Cadbury-Schweppes plc
Birmingham

SPRINGER-SCIENCE+BUSINESS MEDIA, B.V

First edition 1993

Originally published by Chapman & Hall in 1993

Typeset in 10/12pt Times by Falcon Graphic Art, Surrey

ISBN 978-0-7514-0099-1

A catalogue record for this book is available from the British Library

Library of Congress Cataloging-in-Publication data

Food process monitoring systems / edited by A.C. Pinder and G. Godfrey.
-- 1st ed.
p. cm.
Includes bibliographical references and index.
ISBN 978-0-7514-0099-1 ISBN 978-1-4615-2139-6 (eBook)
DOI 10.1007/978-1-4615-2139-6
1. Food industry and trade--Quality control. 2. Food--Analysis.
I. Pinder, A. C., 1953– . II. Godfrey, G.
TP372.5.F65 1993
664'.07--dc20 92-43492
CIP

Preface

The manufacture of foods and beverages is a highly competitive, international industry, and the range of products is becoming increasingly diverse. Manufacturers are required to produce quality foods with the highest possible efficiency and lowest possible cost, and international legislation is imposing strict controls on food safety. Process control is the essential link between quality, safety and cost. Radical changes in the technology of manufacturing bring with them new requirements for monitoring (and ultimately controlling) increasingly complex parameters. The aim of this book is to review the latest developments in monitoring systems, particularly those suitable for the rapid sensing of composition, structure or microbial status. The emphasis is on 'up and coming' methods that have been proven in the laboratory or in other industrial environments, and offer potential in the food sector. As such, it is hoped that this book will increase the general awareness of what new systems have to offer, and will act as a catalyst in the technology transfer process.

The book features chapters on automated machine vision, fluorescence cytometry, infrared spectroscopy, light scattering spectroscopy, ultrasound, mass spectrometry, and chemical and biological sensors. In all cases, the basic approach is to describe the underlying principles, and then to consider the implementation of a particular technique. Examples are given of the practical application to specific problems in the food industry.

It is hoped that this book will appeal to a wide readership, particularly senior management and engineers responsible for product and process development, staff in industrial R&D Laboratories and Research Associations, and members of 'Watch-Dogs' and independent Advisory Boards. It should also appeal to readers from the academic sector in Departments of Food Science and Biotechnology, and in Chemical Engineering, Production Engineering, and Applied Physics Departments with related interests.

I would like to take this opportunity to thank all the contributors for their enthusiasm and effort in compiling this book, especially my co-editor Graham Godfrey. I am also grateful to the publishers for their help and advice. Finally, I am indebted to Rachel Dann for proof reading and help with the index.

Andrew Pinder

Contributors

Professor B.G. Batchelor
School of Electrical, Electronic and Systems Engineering, University of Wales, Cardiff CF1 3YH, UK

Dr J.P. Chan
6 Lothian Crescent, Cyncoed, Cardiff CF3 7HY, UK

Mr S. Gatley
Chemunex S.A., 41 rue du 11 Novembre 1918, 94700 Maisons-Alfort, France

Mr. G. Godfrey
Cadbury-Schweppes plc, Bournville, Birmingham B30 2LU, UK

Dr C. Javanaud
13 Ringwood Close, Little Melton, Norwich, UK

Miss E.K. Kemsley
AFRC Institute of Food Research, Norwich Laboratory, Norwich Research Park, Colney, Norwich NR4 7UA, UK

Dr E. Kress-Rogers
4 Willow Drive, Sendmarsh, Ripley, Surrey GU23 6LF, UK

Professor D. Lloyd
Microbiology Group, School of Pure and Applied Biology, University of Wales College of Cardiff, PO Box 915, Cardiff CF1 3TL, UK

Dr A.C. Pinder
AFRC Institute of Food Research, Norwich Laboratory, Norwich Research Park, Colney, Norwich NR4 7UA, UK

Dr M.M. Robins
AFRC Institute of Food Research, Norwich Laboratory, Norwich Research Park, Colney, Norwich NR4 7UA, UK

Dr M. Wedd
Malvern Instruments Ltd., Malvern, Worcestershire WR14 1AQ, UK

Mr R.H. Wilson
AFRC Institute of Food Research, Norwich Laboratory, Norwich Research Park, Colney, Norwich NR4 7UA, UK

Contents

1 Introduction

G. GODFREY

All manufacturing processes require monitoring and food manufacturing processes are no exception to this. When compared with the chemical process industry, however, the monitoring strategies in the food industry are not in any sense sophisticated and can certainly be improved upon. There are many legitimate reasons for this, and some not so legitimate. To survive and grow in the modern business environment there must be a constant pressure on the costs of manufacture and modern process monitoring systems can be fundamental in achieving this by providing prompt and accurate information upon which decisions can be made by operators, managers and control systems.

Many food manufacturing processes have ill-defined origins (there are in reality very few new foods *per sé*) and the structure and chemistry of many products has, until comparatively recently, been relatively poorly understood. Additionally, the very properties which make many foods unique and attractive have been difficult to define and are still in many cases very difficult to measure objectively.

It is interesting for instance to consider that the characteristics of many foods have their origins in preservation processes designed to store a surplus of perishable raw material for consumption at a later time and that the properties which the consumer has subsequently come to value in these foods are an accidental result of this processing. Some examples include:

- Milk – Butter
  – Cheese
  – Block milk
  – Crumb chocolate
  – Condensed milk
- Meat/Fish – Drying
  – Salting
  – Smoking
  – Canning
- Vegetables – Pickling
  – Drying
  – Canning
- Fruit – Preserving
  – Canning

Freezing has not been included here, because it is a relatively modern process which does succeed in preserving with little change. Canning, however, provides an excellent example of a preservation technique in which many consumers frequently identify the products as 'unprocessed' but which in reality significantly changes the characteristics of the product.

Originally the 'consumer' was just grateful that his winter stores were not rotting, turning rancid, or even growing 'fur', but as adequate food has become assured, flavour and texture have become more important. Now the customer wants the exact consistency and flavour originally conferred by a traditional preserving process (and even in the case of some cheeses, for instance, the original fur) which is now carried out with modern equipment which achieves the same physical objectives but without many of the flavour developments and textural changes caused by the original processes. Thus our interest is transferred from the original monitoring parameter, 'what is the moisture content?' to a much less objective one, 'has sufficient flavour development taken place during drying?' which is linked to the current mode of the food rather than to the original one.

Traditional processes with difficult, labour intensive and often slow analytical procedures, which owed more to technique than technology, have been typical of the food industry and only in about the last ten years has this begun to change rapidly in many areas. Even measurement of basic parameters such as temperature has not been without problems, as the difficult properties of many materials can render even this 'simple' measurement unreliable leading to control systems requiring frequent cross-checking and adjustment of set points, and this type of problem has tended to act as a deterrent to the utilisation of more comprehensive techniques.

The technology of food processing has changed completely in the last twenty years in terms of scale and sophistication, and food manufacture is now a major process industry with all the rights and obligations which that brings. *Effective monitoring of food manufacturing processes is now absolutely fundamental to the safe and efficient operation of plant and equipment and to the production of a consistent, high quality and ultimately trustworthy product for the consumer.* The increasing adoption of well designed and engineered manufacturing processes is transforming the requirements of monitoring from assuring simple parameters to monitoring and ultimately controlling much more complex ones.

Another important factor has been the growth of international trade in food products which has brought with it the need for a structure of dispassionate and independent legislation on food standards. The growth of legislation in particular has moved the food industry to a situation where much higher standards have been required both in terms of composition and safety and also the necessity of having records to demonstrate that standards have been complied with. This has, in fact, been of immense value to the consumer (although national emotions frequently mean that this has gone largely unrecognised) but has inevitably raised costs for the food manufacturing industry. With increasingly sophisticated and capable monitoring systems available it is now becoming possible to drive down these extra costs again and to see further benefits in operational terms in addition to compliance with legislation and compositional requirements.

Without the duality of an efficient operation producing a reliable product there will be a constant drain of a company's resources into higher operating costs, reduced customer satisfaction and poor utilisation of capital equipment. No manufacturer can survive this indefinitely, particularly if he wishes to see products become established, high profit, brands with long term growth (and without growth there is only contraction). It is therefore a basic commercial requirement that monitoring techniques are developed at least at the same rate as the technologies of manufacture. Our ultimate objective must be to increase the consumer's trust in the product, trust which is gained from preference, consistency, safety and ever improving value for money.

As manufacturing equipment becomes more sophisticated, raw materials and services more expensive and customer expectations more acute, then it is only by harnessing and developing the most modern monitoring techniques that we can rise to these challenges and provide what the customer fundamentally (and rightly) demands—a consistent, high quality product at the lowest possible cost.

The food industry has, of course, a number of special problems when process monitoring is considered. The raw materials which have to be processed are natural and therefore by definition variable in their properties and subject to spoilage as well as being complex in composition and structure. In addition the products which are manufactured have to be approved by the most sophisticated and the most subjective monitoring system which will probably ever be devised—the human sensory system. This has a vast range of highly sensitive, interrelated senses with non-linear responses to stimuli. In addition it is uniquely able to respond to different stimuli as a group, and to weight stimuli individually: furthermore, the 'calibration' of this system is highly variable. A further problem is that the physical form of many food products is extremely complex, making effective sampling and monitoring even more difficult, particularly where the human sensory system has a strong response to very low levels of a singular component.

This book examines the principles and application of some of the most modern techniques which are currently making the often difficult journey from the experimental to the routine. Whilst some techniques may already have achieved routine monitoring status in other industries, there remains much to be done in the food industry to identify the ways in which monitoring systems can be applied and how they need to be adapted and

Table 1.1 Advantages of effective food process monitoring—Summary chart

Material	Advantage of rapid and effective monitoring
Raw material before delivery	Assured continuity of production Reduced stockholding Reduced costs of return of goods etc. Time to source alternatives before production requirements become critical
Raw material at warehouse	Efficient production planning and plant scheduling Early adaption of process to natural variations Assurance of delivery to customers Time for decision making
Material in process	Potential for improved control—reduced time lags etc. Early warning of deviations allows planned correction Reduced scrap losses of material Reduced operating costs—less reprocessing Higher labour productivity and plant availability Quality critical parameters maintained closer to optimum improving overall quality and product assurance Improved customer confidence Reduced product in process by more rapid clearance
Finished product	Improved product quality and consistency Increased customer acceptability and confidence Reduced complaints/rejections Better understanding of quality critical parameters Reduced stockholding by more rapid clearance Better response to marketing requirements

developed. An example is infrared spectroscopy which is routinely used in the chemical industry for on-line analysis of gas mixtures but is in its infancy in the food industry. This is not solely because of the complex nature of the task which it is being asked to do in the food industry—frequently to identify a relatively poorly defined species against a heterogeneous and often variable background—but also because of the difficulties of sample presentation and confusion about which control variable should be operated upon. Without an identified control variable, monitoring is, after all, utterly futile.

However, this book is concerned with the monitoring of process operations and does not consider control. Control, based on the output from sensing and monitoring systems of all types, is an important subject of its own and it would be at least confusing to try to cover both subjects in one book. It is perhaps worth mentioning, however, that control system technology is developing at least as rapidly as monitoring technology and that in order to exploit the capabilities of either to the full a degree of integration is required. Modern control techniques are important to monitoring technology because of analogies between gathering and interpreting a number of data streams for control purposes and interpolating complex sensor input to the parameter being measured.

An important factor which has frequently hindered the adoption of better monitoring techniques is the relatively low financial margins enjoyed by the food process industry compared with, for instance, the fine chemical and pharmaceutical industries. This is not an unreasonable comparison as the levels of reliability and hygiene which have to be achieved are not inconsistent with each other. This is a factor which to some degree will be self-correcting. In order to be competitive in the modern food manufacturing environment, processes will inevitably become more capital intensive and as a direct consequence sufficient margin will have to be generated to permit proper monitoring, as only by this means can large scale operation become adequately efficient to repay the investment.

It is not only the capital intensive nature of the plant which will force the adoption of more advanced monitoring systems. Most companies have substantial amounts of working capital tied up in raw and part processed materials and finished goods. This can often be substantially reduced by increasing the certainty of operation and reducing the time associated with final quality control (QC) clearance. For example, in the increasingly universal practice of positive microbiological clearance for finished goods, if the time before release can be reduced by a relatively modest amount the commercial benefit can be immense. For example:

- Original time for clearance — 5 days
- Value of 'held' stock — £1 000 000
- New time for clearance — 3 days
- Value of 'held' stock — £600 000
- Reduction in working capital — £400 000

This reduction in working capital can be reflected in either reduced borrowing or in an equal investment in manufacturing capacity. In either case the benefit to the company will be of the order of £100 000 per year, before such advantages as longer shelf life, reduced warehousing cost, improved response to manufacturing disruption and so on are taken into account.

Stocks of raw materials can be nearly as expensive to finance as finished goods and as all manufacturers strive to reduce stocks it becomes increasingly important to be able to quantify the important quality aspects of raw materials as rapidly as possible in order to avoid production down time due to rejection of raw materials. This is particularly important where fundamental raw materials are imported and, even more so, perishable. In this case the alternative to acceptance can easily be a loss of production with little redress being available to the processor. New developments discussed here should help to move towards obtaining instant or near instant results once key objective parameters have been agreed with suppliers and to give irrefutable evidence of, for instance, a pesticide residue problem before the material is accepted from the supplier, or even whilst it is still in transit.

Development of monitoring techniques which are rapid enough to provide quality assurance within transit times will allow food manufacturers to move towards the low costs and flexibility associated with 'Just in Time' techniques beginning to be applied in the engineering industry in particular, as well as some of the more advanced members of the retail trade. Pressure for rapid monitoring techniques is also a natural result of shorter shelf life and the consumers' preference for 'lighter' levels of preservation with ever higher degrees of safety and hygiene. As the shelf life shortens so the time available for post-manufacture procedures must also reduce in order to give an acceptable in-store sales life.

Fortunately (or otherwise, depending on your point of view) the environment is changing rapidly to accommodate these factors, and will ultimately eliminate or at least marginalise those companies not operating at a high level of efficiency. The ISO9000 standard and national equivalents are also placing demands on manufacturing operations for monitoring and documentation which cannot be ignored and which can in many cases only be complied with by exploiting efficient modern monitoring systems.

Another example of change is the increased effectiveness of tracing the source of microbiological contamination back to its origin. This means that a company simply has nowhere to hide if it is responsible for an incident of microbiological contamination and therefore must take the most stringent precautions, as increasingly the penalty for such an error is not just a punitive fine. There are several recent cases of commercial extinction due to loss of consumer and customer confidence following microbiological and other types of contamination.

Food process monitoring can be considered to cover two quite different types of monitoring. The first and most obvious is the measurement and control of fundamental physical parameters, such as temperature, pressure, and so on. These will frequently have a major effect on the ultimate quality characteristics and should not be forgotten as basic control parameters, but increasingly frequently, where finished goods manufacture is involved, it is now necessary to monitor directly the ultimate parameter which it is hoped to control and it is this aspect which is particularly addressed in this book.

A useful example of this occurs in many pasteurisation processes where the correct temperature/time relationship may be controlled and monitored with conventional techniques, but the ultimate parameter, the remaining levels of viable organisms, must often be monitored before the product is released for sale. This means that rapid microbiological methods are of increasing importance and each succeeding hour by which the time necessary for a procedure can be reduced is almost exponentially valuable. In this context the techniques of flow cytometry may have much to offer and the principles and developments available are considered in chapter 2.

Microbiological quality monitoring is perhaps the most rapidly growing and one of the most expensive areas of product monitoring. It is likely that in the foreseeable future (at least in developed market places) all food products and many raw materials will require positive clearance before sale.

Ultrasonic analytical techniques are developing rapidly as understanding is gained of the ways in which the structure and composition of foods alter the ultrasonic properties. This includes not just simple attenuation but frequency and phase specific properties which are able to give important results.

The ability of ultrasound to respond to structural and certain physical properties in foods and raw materials places it in an almost unique position, as does the relative robustness of sensor technologies. Excellent examples of this include monitoring of the condition of emulsions, the viscous and glassy nature of products and the presence of parasites in fish tissue.

Ultrasound characterisation will be an important technique developing as the need for monitoring of the structural properties of complex foods increases in importance and the discussion of the subject in chapter 6 will be invaluable to those new to the subject and also to those with some experience of the technique.

A rapidly emerging and powerful new area is the development of systems which will visualise (not merely in the human 'visual' spectrum) and analyse products and raw materials. This allows a system to be designed which will fulfil a variety of roles, for example:

- Detecting items with a missing element—for example the cherry on top of a cake or some other decorative feature.
- Detecting items which are incorrectly shaped and which might be outside the tolerances of a high speed packaging line.
- Detecting incorrect colour due to the wrong level of bake or some similar variable.
- Detecting the fill level in containers which may be opaque in the visual spectrum but which may be interrogated by some other frequency.
- Detecting the presence of foreign bodies such as glass or plastic which might be hidden in the bulk of material.

A particularly important aspect of these systems is that they allow us a greater degree of certainty against infrequent occurrences than does manual inspection. If 99.9% of output of a product has the correct physical characteristics (a fairly modest objective) then the chances of a human operator spotting the exceptions are not good, because he or she will simply become fatigued with examining good product before a faulty one comes along. On a confectionery bar line, for example, throughputs of twenty thousand units per hour would be modest and to expect a human operator to identify and pick out of the order of a hundred malformed units is barely reasonable; the inspection process will be at best haphazard.

The last mentioned area of foreign body detection is a particularly vital one as the need for increasing levels of surety in products becomes ever more important, whilst the increased use of levels of mechanical handling reduces the effective degree of human inspection of raw materials. In addition there is an increasing need for manufacturers to be able to demonstrate intrinsically safe systems in the face of increasing litigation and incidents of deliberate contamination of products at the point of sale by the criminal and the deranged.

The ideas presented in chapter 4 should encourage the development of new control ideas where particular problems are amenable to this type of solution.

Many food products and raw materials depend critically for their properties on particle size. Even in recent years particle sizing was such a laborious task, generally based upon sieve analysis and in a few cases on sedimentation, that for most purposes relatively simple levels of analysis were employed, frequently based on small samples. Modern developments have transformed this situation, allowing rapid analysis of statistically relevant samples to give complete information.

In addition modern techniques have allowed the examination in much greater detail of the droplet sizes in emulsions and the investigation of the roles of very fine fractions of particles. This is important because the fine fractions, whilst small in terms of mass fraction, are predominant in terms of number and in the fraction of interphase contact area which they represent. The true importance of this type of information to the properties of many products is only starting to become clear as analytical techniques increase our understanding of structure which in turn enhances the need for particle size monitoring in order to control and optimise particle size distributions.

An interesting example is the chocolate industry where originally particle sizing was carried out using a variety of solvent based sieving and sedimentation techniques which gave a reasonable assurance of the product being within a set organoleptic specification. These techniques have lapsed in use over the years because they are labour intensive, require significant skill and also utilise solvents which are not regarded with favour in modern food factories.

The replacement was, in many cases, the use of a digital micrometer to attempt a measurement of a sample dispersed in vegetable oil. A dispassionate examination of this technique will rapidly suggest that it has many weaknesses and, in all but the most skilled and reputable hands, will give whatever answer the operator wants it to give. A further weakness is the tiny sample size involved, of the order of milligrams.

The problem of chocolate particle-sizing has been transformed by the introduction of laser scattering equipment. This technique gives an excellent statistical view of the particle size distribution and has been widely adopted, but much better analytical design could have been achieved if more consideration had been given to some of the ideas presented in chapter 5. This particularly applies to criteria for minimum sample size and number of measurements, effect of particle shape on readings and on what basis 'size' should be determined; number, length, area, volume and so on, all of which may relate to different aspects of behaviour.

If the modern techniques, frequently based upon analysis of large numbers of particles and giving a wealth of statistical information, have a weakness, it is in their lack of capacity to reduce the data to something directly usable but which still exploits the information available. Too often an operator (and often a senior analyst) will use one number from a page of data because no generalised analysis of the data has been undertaken when the technique was developed.

Chapter 5 on particle sizing introduces a range of available techniques and should give much useful advice to those involved with finely divided materials.

A most innovative field of monitoring technique development is that of chemical sensors. Until recently even simple techniques such as pH measurement have rarely been applied successfully outside the laboratory, due not only to the difficult physical conditions in the production environment but also because process designers have been understandably reluctant to permit glass based electrodes to be installed directly in equipment.

Several areas of development have combined to give the opportunity for sophisticated chemical sensing in the solid, liquid and gaseous phases:

- Improvements in the understanding of the chemistry and biochemistry of both raw materials and finished products have allowed key entities to be identified and their relationship with organoleptic properties to be elucidated.
- Developments in materials—particularly porous but mechanically strong polymers—have allowed sensors to become physically robust enough to be installed without danger of product contamination.
- Novel sensor techniques have allowed sensing elements which are sensitive to specific entities and which can also tolerate the process environment for long periods to be designed.
- Advances in signal processing have allowed much noisier and inferential data to be processed into useful information. In particular neural networks have permitted several data sources to be combined to give a provable conclusion without the rigour of a strict mathematical model being available.

Because it has the potential to answer specific questions, particularly by developing 'designer sensors' to respond directly to key entities which may be difficult to elucidate by conventional means in the laboratory, this technique has a major contribution to make to the food manufacturing industry. In particular there exists the opportunity rapidly to quantify in an unarguable manner many parameters which have hitherto been essentially subjective and thus unavailable to a control system in a routine and objective form.

Chapter 8 on chemical sensors provides a view of the development of process acceptable sensors for relatively conventional parameters such as pH and the possibilities now becoming available in such fields as freshness testing of meat and fish and aroma analysis using arrays of sensors.

An analytical technique which can be considered to be in a similar family to chemical sensors is infrared spectroscopy. In this case rather than detecting the presence of complete chemical identities the technique identifies and quantifies the frequencies associated with particular molecular bonds within a molecule by absorption from an incident beam of near infrared (NIR) or mid infrared (MIR).

The advancement of this technique has hinged on the development of improved detector technology as well as sample presentation and signal processing developments, particularly as Fourier transform (FT) infrared analysis has evolved. In addition to the obvious uses in monitoring the composition of gas mixtures, where it has been used in the process industry for many years, the technique is now being used on liquids in flow cells and also by reflection on solids and semi-solids.

Because of the highly derived nature of the signal, the successful application of the technique does depend quite heavily on calibration methods and must be seen as an inferential rather than an absolute measurement system. The exposition of the principles involved in chapter 3 should be of assistance in identifying new areas where the technique may be developed into a useful method as well as highlighting some existing applications.

Moving further into the area of analytical techniques which are being developed towards use in monitoring leads to mass spectrometry. Similar to chemical sensing in its dependency on the identification of the role in foods of individual species, it is ultimately capable of detection of such species in minute quantities and is thus particularly useful in flavour analysis.

The detection of pesticide and drug residues in foodstuffs is a another particularly significant use of mass spectroscopy and one in which its sensitivity and selectivity have few rivals. However, it is by its nature a capital and skill intensive technique which will be more readily found in the research rather than the QC laboratory for some years to come.

Another particularly interesting field is pyrolysis mass spectroscopy which allows the rapid and specific identification of microbiological organisms. This may be a major area of development as the need for microbiological 'detective work' rapidly to identify specific sources of contamination becomes increasingly important.

In many ways mass spectroscopy can best be considered a developing field of immense potential rather than one in which techniques exist and the discussion in chapter 7 of the principles and some of the techniques involved will help to lead to further practical developments.

2 Fluorescence cytometry for the rapid analysis of food microorganisms

A.C. PINDER and S. GATLEY

2.1 Introduction

A key factor in the implementation of active product quality management in the food industry is the delay associated with microbiological analysis. Incoming raw materials need to be inspected before acceptance, production processes such as fermenters need to be continuously monitored, final products need to be tested for quality, and environmental cleaning procedures need to be assessed for effectiveness. Any delay in identifying batch failure during production, due for example to contamination or uncontrolled changes in manufacturing conditions, may result in the loss of expensive raw materials and a consequent reduction in plant efficiency. Microbiological data on the quality of the final product (particularly if it is a short-life product) may not be available until several days after that product has been released on to the market; a bad report will mean product recall, with all its associated cost implications, and will lead to a reduction in customer confidence.

For many years, microscopy and microbial plate counts have formed the cornerstone of industrial microbiology. The primary limitation of these classical microbiological techniques arises from the need to grow single cells into colonies of large numbers of cells before they can be detected. For example, if an organism has a mean doubling time of 30 minutes, it will take between 18 and 24 hours to develop a colony that is visible to the naked eye (10^6 organisms). With common yeast spoilage organisms, this period will be extended to three to seven days.

There has been an intense effort to develop alternative 'rapid methods' for the detection of microorganisms that overcome the limitations of classical techniques. Many of these are based on the measurement of a signal that is 'global' in nature, for example the variation in electrical impedance of a liquid sample. Such global measurements reflect the sum total of all organisms in the sample and therefore give only very limited information. The measurements are prone to errors and artefacts, and internal cross-checking is problematic. In mixed populations, it is difficult to differentiate the signal generated by one organism from those generated by other organisms present. This is particularly a problem in the food industry; for example, in fermented milk products small numbers of spoilage yeasts must be detected against a large background of *Lactobacilli* and *Streptococci*.

The ideal rapid method should therefore provide information on the activity of one or more particular organisms in the presence or absence of others. Kinetic data may be important, since cell activity will vary with time in response to the inhibiting effects of antibiotics or other agents. Furthermore, the behaviour of individual cells within populations is also often important, as mutations and changes can effect the population over time. In the search for new methods, techniques based on the optical detection of individual cells on a one-by-one basis are especially promising. In combination with fluorescent reagents for marking particular biological properties, they offer the opportunity to identify highly specific organisms present even in very low numbers.

2.2 Optical analysis

Optical cytometry techniques owe their origins to classical light microscopy. The fluorescence microscope can provide remarkable resolution, chemical specificity and sensitivity; it can detect and map substances rendered visible by fluorescent stains specific for particular constituents (Munck, 1989a; Rost, 1992). However, microscopy is tedious and time consuming, as each field is selected and viewed manually by a skilled operator. Techniques have been developed in which the sample is viewed and assessed automatically, either by moving the sample in a continuously flowing liquid stream past the measurement system (flow cytometry), or by mechanically scanning the measurement system over a fixed sample (confocal scanning cytometry) (Gatley, 1990).

2.2.1 Flow cytometry

Flow cytometry as a technique was developed over twenty years ago and has made a significant contribution to biomedical research and the study of eukaryotic cells (Mackenzie and Pinder 1987; Melamed *et al.*, 1990; Shapiro, 1988). Its application to the study of yeasts and microbes, particularly in an industrial environment, is, by comparison, much more recent. The size, mass, nucleic acid and protein content of a bacterium are typically one thousand times smaller than a mammalian cell, and measurement of these parameters requires an instrument with the highest optical resolution (Pinder *et al.*, 1990; Steen *et al.*, 1982).

The basic principle of flow cytometry is that the sample is forced to flow in a narrow, well-defined liquid stream through a fixed optical excitation/detection system. Thus, individual organisms in the sample pass one at a time through the optical 'interrogation' zone. Strategically placed photosensors measure optical changes (which are characteristic of cell type) as each cell passes. Signals from the sensors are electronically processed and analysed by computer.

Figure 2.1 is a schematic diagram of a typical flow cytometer (although it should be noted that there is some variation in precise details between instruments from different manufacturers). The sample stream needs to be positioned precisely on the optical axis of the cytometer, and this is accomplished by hydrodynamic focusing (Pinkel and Stovel, 1985). A jet of biologically compatible fluid, referred to as the 'sheath' (e.g. buffered saline), is forced through a small orifice to form a very fine stream. The orifice is typically 50–200 μm in diameter, and the driving pressure is carefully adjusted so that the flow is laminar. Flow rates are usually between 1 and 10 ms^{-1}. The liquid sample is then slowly introduced through a tube into the centre of the sheath, where the velocity gradient is such that the cells in the sample are accurately centred and pass one at a time through the orifice.

A short distance below the orifice the cell stream encounters the interrogating light beam (in this case, from a laser). The beam is focused to a spot between 20 and 50 μm across. As each cell passes through this spot, a short (~5 μs) pulse of light is emitted which is captured and imaged by lenses onto photosensors (usually photomultipliers).

The instrument shown in Figure 2.1 has an arrangement of four photosensors. One detector, collinear with the interrogating beam, measures light scattered forward at narrow angles (< 15°) as each organism passes in front of the laser. (The intense laser beam itself is prevented from reaching the detector by an external obscurator or beam-stop.) This forward angle light scatter (FALS) signal is related to the size of the cell

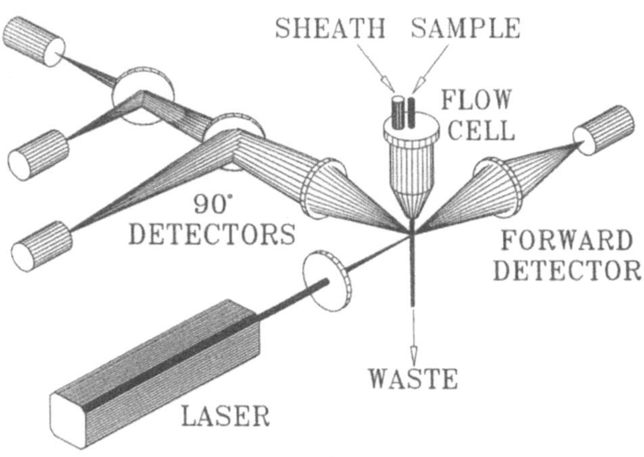

Figure 2.1 Schematic diagram of a typical flow cytometer.

(Mullaney *et al.*, 1969). A FALS signal alone is therefore sufficient to discriminate, for example, between bacteria, yeasts and larger cells. A second light scatter detector, placed at 90° to the principal optical axis of the instrument, measures refracted light, which is related to the granularity of the cell (Salzman *et al.*, 1975). The utility of the cytometer is, however, greatly enhanced by the presence of further detectors, tuned with optical filters to detect fluorescent light emissions over specific ranges of wavelengths. These permit the detection of a great wealth of information about cellular characteristics, through the use of specific fluorescent reagents (see section 2.4.1). With appropriate combinations of dichroic beam-splitters, it is possible, for example, to distinguish between green, orange and red fluorescent emissions from three specific stains, all of which are excited by a light source in the blue region of the spectrum.

Electrical pulses from the various detectors in the flow cytometer are shaped and amplified by dedicated electronic circuits (Hiebert and Sweet, 1985). Depending on the exact optical arrangement, the pulses may be classified by detecting the peak height, or by integrating over the 'time of flight' of a cell through the interrogation zone. In the latter case, it may even be possible to extract information about the distribution of fluorescence within an individual cell, rather than averaged over the whole.

Ultimately, the electronic signals are digitized and fed to a dedicated microcomputer for statistical analysis (Dean, 1990). The operator can select the most suitable form in which to present the data. This may be as uncorrelated, single-parameter histograms, showing for instance the distribution of cell-size or fluorescence over the total population. Alternatively, data may be presented as a correlated, multi-parameter histogram, displayed as an isometric or contour plot of a solid surface. Statistical information can then be obtained on, say, a sub-population of cells which is 'large' in size but 'weak' in fluorescence, compared to another which is of 'small' size but 'strong' fluorescence. Programming the computer to select logical combinations of measurement parameters allows differentiation of even two very similar sub-populations of cells.

Flow cytometers can analyse samples at flow rates of up to about 0.5 ml min^{-1} and cell count rates of up to 10 000 s^{-1}. There is a good correlation between classical culture techniques and flow analysis of samples of between 10^7 cells ml^{-1} to 10^2 cells ml^{-1}.

2.2.2 Confocal scanning cytometry

Whilst flow cytometry is ideal for population analysis, the detection of very low numbers of cells (often referred to as 'rare event analysis') requires a culture enrichment step prior to analysis. For sterility testing, the detection of single microbial cells may be required in large sample volumes. Since sterility testing is often carried out using filtration of the sample, it would be convenient to scan the filter surface for microbes directly.

The direct epifluorescence filter technique (DEFT) counts fluorescently-stained bacteria on the surface of a membrane filter, using a conventional fluorescence microscope (Pettipher, 1983). Counting the organisms by eye is tedious and time consuming, and even with a semi-automated image analysis system the practical lower limit is over 10^3 colony forming units (cfu) g^{-1} (Rodrigues and Kroll, 1988). In the same way that confocal scanning optical microscopy was developed for improved images compared to conventional light microscopy (Wilson, 1990; Wilson and Sheppard, 1984), so confocal scanning cytometry is now being developed to provide a new technique for the detection of microbial cells on membrane filter surfaces (Gatley, 1991; Kroll *et al.*, 1989).

The essential elements of a confocal scanning cytometer are illustrated in Figure 2.2. The input light beam (from a laser) is scanned across the fixed filter membrane by two galvanometer-driven mirrors. The laser beam is focused down to a spot of between 1 and 20 μm in diameter at the surface of the membrane. This results in a considerably higher sample irradiance than can be achieved using full-field illumination, which is particularly important in detecting weak fluorescent signals.

Reflected and emitted light from the spot on the sample is imaged onto a low light level detector (photomultiplier or intensified charge-coupled device, CCD). Unlike conventional imaging systems, spot scanning does not mean compromising high resolution for a wide field of view. A conventional imaging system provides information from the entire object field simultaneously and the achievable resolution is a function of the

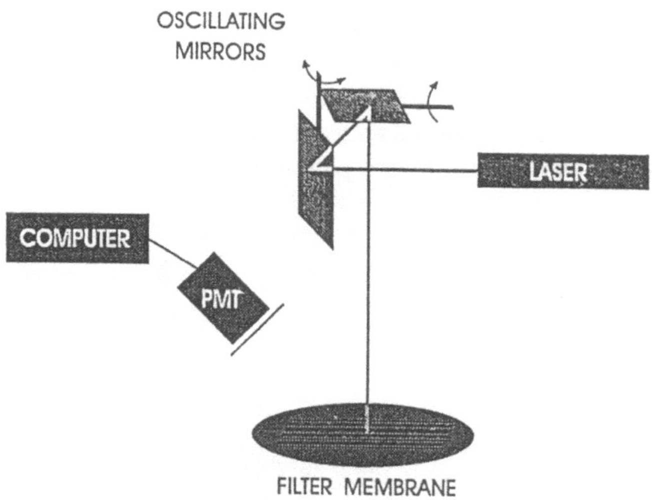

Figure 2.2 Schematic diagram of a typical confocal scanning cytometer.

optical components: it is very difficult to maintain a high resolving power over a wide field of view. However, with a scanning system, even with very large fields, only a small part of the object is viewed at any one time and the optics need be designed purely to provide detection of signals from a single point on the membrane.

The second key feature of the instrument is that the detector is confocal with the scanning spot. That is to say, the field of view of the detector is matched to the area of illumination; all light from outside the focal point is rejected. This results in a high signal-to-noise.

In a standard confocal scanning microscope, the detected signal from each spot on the sample is acquired by a computer and used to build up an image of the whole field in a video framestore. In theory, the same system could be used for cytometry, with special image processing algorithms to detect and measure key features of the image held in the framestore. In practice this would be very inefficient, requiring large amounts of memory, processing power and execution time. Typically, the system is required to scan a filter membrane 20 mm in diameter; statistical sampling is not possible if rare events are not to be missed. Ideally the analysis should be kinematic, that is it should be possible to compare scans taken over a period of time in order to investigate growth in the sample. The solution employed in the confocal scanning cytometer is to process the detector signals as they are acquired. When a cell is detected, its signal is recorded alongside the position of the scanning laser spot. In this way a digital map is built up of the surface of the membrane. Subsequent scans can then easily be overlaid for comparison.

As with flow cytometry, multi-parameter detection is possible, using multiple detectors and appropriate combinations of optical filters.

2.3 Light sources for optical cytometry

The light source in an optical cytometer must be chosen to excite the required fluorochrome(s) with maximum efficiency. It is usually either a laser or an arc lamp. There are distinct advantages and disadvantages to both (Wheeless and Kay, 1985). Lasers generate an intense, stable and spectrally pure light beam, which may be focused down to a spot of the order of micrometres in diameter. The optical geometry of a laser system can be kept relatively simple, but the overall cost of the instrument will be high (in some cases, very high). Arc lamps are cheaper and emit over a broader range of wavelengths, but are less stable. Although the arc lamp is, in itself, a relatively simple source of light, an optical system of greater complexity than with a laser is required to make efficient use of the available light. Consequently, current practice is to use a laser in applications where high sensitivity scanning of morphological features in the sample is required, but to use an arc source if large aperture excitation and detection optics can be employed. (Incandescent filament lamps are not generally suitable for reasons of output power and spectral distribution.)

The mercury arc lamp is a familiar source for conventional light microscopy. It has a line spectrum spanning the region 240–600 nm, with principal emissions at 365, 405, 435, 546 and 578 nm (Figure 2.3(a)). A xenon arc may be preferred at other wavelengths, since it has a continuum spectrum from 190 nm to 2.6 μm, with peaks between 750 nm and 1.0 μm. Miniature arc lamps rated at 100 W are usually sufficient, operated from a simple direct current (DC) supply and cooled by natural air convection. The distribution of irradiance from the arc is highly non-uniform, centred on a 'hot spot' at the cathode. Care must therefore be taken in the design of the illumination optics, and fading or instability in output may arise due to cathode ageing.

Argon ion lasers are widely used in research-grade laboratory cytometers. They can emit at ten discrete wavelengths in the range 454–529 nm, the brightest and most commonly used being 488 and 514 nm (Figure 2.3 (b)). Some high power lasers can also operate over a narrow range of the ultraviolet (UV) spectrum, from 330–365 nm. A nominal '5 W all-lines' laser will produce about 1.5 W at 488 nm and 0.4 W in the UV, but requires 25 kW of electrical power to drive it and over two gallons per minute of circulating water to keep it cool. Such sophisticated technology is expensive to buy and expensive to run, though reliable operation over a relatively long lifetime should be expected. (Krypton ion and ion-pumped dye systems are also sometimes used, extending the range of lasers into the yellow to red spectral region, but these highly elaborate instruments will not be discussed further in this chapter.)

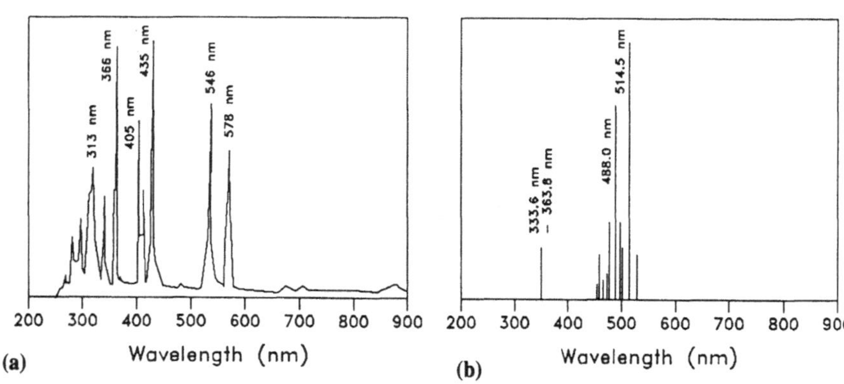

Figure 2.3 Emission spectra of two commonly used light sources for flow cytometry: (a) mercury arc lamp and (b) argon ion laser. The vertical axis in each case is arbitrary, but gives an approximate indication of the relative spectral distribution of intensity. (The argon ion emission shown as a discrete line in the UV is, in fact, a narrow band of wavelengths in the range 336.6–363.8 nm.)

The water-cooled ion laser is not suited to a general industrial environment. Smaller, much cheaper, but less flexible, air-cooled argon ion lasers are. They operate from a standard single-phase mains supply (2 kW) and are fan cooled. Fewer optical wavelengths are available, with the power at 488 nm reduced to 100 mW maximum, and no output at all in the UV. It is advisable to leave a 'warm-up' period of about half an hour after starting, to guarantee output stability.

Another air-cooled device, the helium–cadmium laser, offers up to 50 mW of light at 442 nm and 10 mW at 325 nm. It may be suitable for use with some UV-excitable DNA stains, but owing to the comparatively low power output its use for cytometry has not been widespread. Also, the simple helium–neon laser is of limited applicability: it has usable optical power at only one wavelength (maximum of 50 mW at 633 nm). Although some antibody labels are available which can be excited at this wavelength, there are no reliable DNA or viability stains. The helium–neon laser may be worth considering for light scatter measurements alone or for some specialized applications, or in combination with an air-cooled argon ion laser for two-colour excitation.

The advantages attendant with lasers are monochromatic, highly polarized, highly collimated beams with uniform, stable distribution of optical power over small diameters (~ 2 mm). A laser beam can be focused to 'diffraction limited' dimensions (i.e. the size of a single organism or less).

For immunofluorescence analysis, a laser source is to be preferred because of its greater intensity. However, with DNA stains, when many (orders of magnitude) more photons are emitted, a mercury arc may well be adequate. In this case, the availability of UV wavelengths is an added bonus. With a laser source, the temptation is to keep increasing the power in order to increase the sensitivity of detection. This has limited returns: a saturation level is soon reached, and irreversible photo-bleaching of the fluorophores will start to occur. A far more effective route to enhanced sensitivity lies in the use of high efficiency optics for collecting light emissions, sharply-defined spectrum-limiting optical filters for discriminating dye fluorescence and sensitive detectors (photomultipliers or intensified CCDs) for low-noise signals, together with fluorescent staining techniques giving a high yield of photons.

The design of a system for optical cytometry is highly dependent on the light source to be used. The choice of light source is critically dependent on the potential application and fluorochromes required. In turn, the efficiencies of the fluorochromes will impose constraints on the instrument. These three key factors must be considered together and not in isolation. Fluorescent labelling procedures have evolved from classical microscopy, with blue/green wavelengths very much in mind, favouring argon ion laser sources. The light source of the future will probably be the solid state laser. Red-emitting devices are already in use for optical communications and domestic compact disc players, for example. Lower wavelength devices are under development, but the potential cost advantage has yet to be realized.

2.4 Parameters measurable by optical techniques

The strength of optical cytometry techniques lies in their ability to measure a diverse range of cellular parameters, often simultaneously, on a cell-by-cell basis (Table 2.1). Some of these parameters can be measured without staining, while others require the use of highly specific fluorescent reagents.

The use of forward angle light scatter and 90° light scatter to measure cell size (cross-sectional area) and cytoplasmic granularity, respectively, has already been mentioned. The autofluorescence of some cells is a useful measure of cell content. Pigments such as lipofuscins, haemoglobin or chlorophyll give characteristic signals, as do pyrimidine and flavin nucleotides (Olson *et al.*, 1983; Thorell, 1981).

2.4.1 Fluorescent labelling techniques

Though much can be gleaned from untreated cells, it is through the use of fluorescent dyes or 'labels' that the full potential of optical cytometry is realized (Munck, 1989b; Waggoner, 1990). As the techniques become established, the range of problems to which they can be applied are expanding dramatically. Cell staining techniques already well known in analytical cytology and microscopy have been adapted, and other highly specific dyes and labels have been specially developed.

Some of the ways in which an organism may be labelled and rendered visible to a fluorescence detector are shown diagrammatically in Figure

Table 2.1 Parameters measurable by optical cytometry

Structural	Functional
Cell size	Redox state
Cell shape	Enzyme activity
Cytoplasmic granularity	Endocytosis
Pigment content	Membrane integrity/permeability
DNA content	Membrane fluidity
RNA content	Membrane potential
Chromatin structure	Surface receptors
Surface sugars	Intracellular receptors
Total protein	Intracellular pH
Basic protein	DNA synthesis
	Membrane-bound Ca^{2+}
	Intracellular Ca^{2+}

2.4, and the fluorochromes most commonly used are listed in Table 2.2.

Fluorescent labels offer a unique advantage for optical cytometry, namely that they are excited by light at one wavelength but re-emit light at a higher wavelength (Guilbault, 1989). Thus, using a combination of labels with non-overlapping emission spectra, it is possible to measure several cellular parameters simultaneously (e.g. real time analysis of growth curves, with viable counts of specific target organisms in mixed populations).

2.4.1.1 Nucleic acid stains. There are a number of situations in the food processing industry where a simple measurement of total bacterial concentration is important, and speciation or viability testing is, in the first instance, not required. In such situations the use of a fluorescent dye that binds directly to DNA and/or RNA may be appropriate (Pinder *et al.*, 1990; Patchett *et al.*, 1991). There are many such dyes to choose from (Latt and Langlois, 1990). Ethidium bromide, acridine orange, and propidium iodide, for example, bind to DNA and to double stranded RNA. These stains can be made specific for DNA by the addition of an RNase to the sample. Other stains are more selective in the way they bind to DNA. One group, that includes DAPI (4, 6 diamidino-2-phenylindole) and the Hoechst compounds 33258 and 33342, binds preferentially to adenine–thymine regions; another group, including chromomycin, mithramycin and oligomycin, binds to cytosine–guanosine regions (van Dilla *et al.*, 1983). Of these stains, only Hoechst 33342 does not intercalate with the DNA, and so does not disrupt its structure: cell viability is retained.

2.4.1.2 Fluorescent labels for antibodies. Whilst it is possible to some limited extent to differentiate between different types of organism solely on the basis of their relative size and DNA content (see later), true

Table 2.2 Fluorescence probes for optical cytometry

Nucleic acid probes	Antibody labels	Viability probes
Ethidium bromide	Fluorescein isothiocyanate (FITC)	Propidium iodide
Propidium iodide	Tetramethylrhodamine isothiocyanate (TRITC)	Acridine orange
Acridine orange	Rhodamine isothiocyanate (XRITC)	Rhodamine 123
Diamidinophenylindole (DAPI)	Texas red[a]	Fluorescein diacetate (FDA)
Hoechst 33342	Phycoerythrin	ChemChrome[b]
Hoechst 33258	Phycocyanin	
Mithramycin	Allophycocyanin	
Oligomycin		
Chromomycin		

[a] Texas red is a trademark of Molecular Probes Inc.
[b] ChemChrome is a trademark of Chemunex S.A.

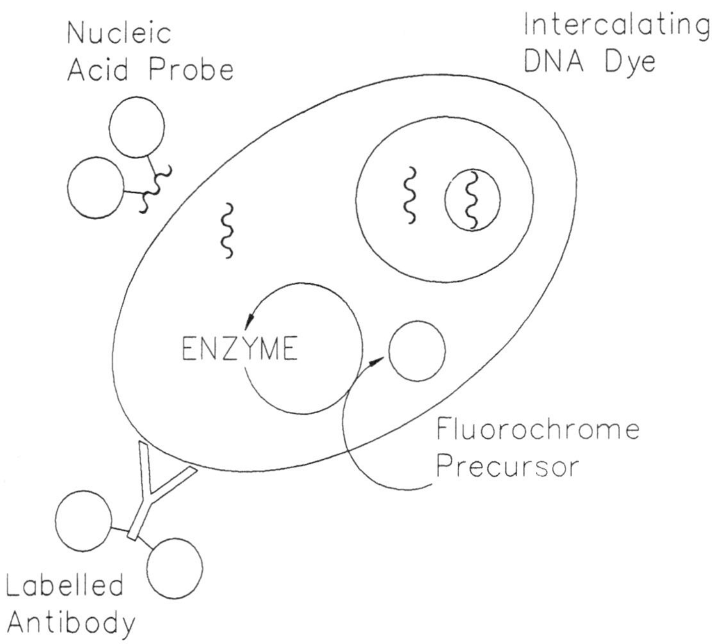

Figure 2.4 Schematic representation of methods by which cells may be fluorescently labelled.

speciation requires a more sophisticated technique. Immunological methods have been developed that use antibodies to detect antigens on the surface of an organism. Highly uniform 'monoclonal' antibodies (MAbs) can be tailored to recognize an antigen binding site as unique as a specific conformation of five or six amino acid side chains on a protein, or of five or six sugar residues on a polysaccharide (Kohler and Milstein, 1975). For use in food process monitoring, antibodies can be raised against specific pathogenic or spoilage organisms (yeasts or bacteria). The specificity and avidity of the antibody–antigen reaction, when properly designed, offers a highly specific, sensitive technique. The antigens may be measured *in situ* without the need for isolation, and interference from the product matrix can be minimized. The conjugation of antibodies with fluorescent dye molecules, and the use of a detection system based on optical cytometry, results in an assay that is relatively rapid, robust and simple to perform.

Some of the most widely used fluorescent labels are listed in Table 2.2. Of these, fluorescein isothiocyanate (FITC) is the most popular: it is excited by the blue 488 nm wavelength from an argon ion laser, it has a high quantum efficiency with emission around 520 nm (green), and conjugation procedures are well established (Haughland, 1983).

If it is required that two different antibodies be used at the same time (two-colour immunofluorescence), they may be labelled respectively with

FITC and the rhodamine derivative XRITC, or Texas red (Dean and Pinkel, 1978). These latter two are excited around 570 nm (orange) and emit at 590–630 nm (red), so there is no 'crosstalk' with the FITC signal. The blue and orange excitation wavelengths are not both available from the same laser source, so two lasers are required, adding considerably to the cost and complexity of the detection instrument.

Fortunately, an important new class of dyes have been developed based on phycobiliproteins found in fluorescent algae (Oi *et al.*, 1982). These include phycoerythrin, phycocyanin and allophycocyanin. Using FITC in combinations with derivatives of these dyes, it is now possible to perform two- or three-colour immunofluorescence measurements with the same excitation source (Festin *et al.*, 1990; Russ, 1990).

There are essentially three methods for fluorescently labelling an antibody: indirect, avidin–biotin, and direct (see Figure 2.5) (Goding, 1986). With the indirect method, two antibodies are used. The first antibody is unlabelled and is made to target the specific antigen of interest. The fluorescent label is attached to the second antibody, which in its turn binds to antigens on the first antibody. An advantage of this method is simplicity: the primary antibody, which may well be highly specific to a particular organism and manufacturing process, does not need to be conjugated directly with a fluorophore. Secondary antibodies against primary targets are readily available and may be bought-in pre-labelled. The disadvantage of this method is that it cannot be used for two-colour detection of two antibodies of the same immunoglobulin class.

A compromise approach to fluorescent labelling uses the avidin–biotin system (Bayer and Wilchek, 1980). First, the antibody is biotinylated by reaction with NHS–Biotin, liberating free NHS (N-hydroxysuccimide) and leaving the biotin bound to the antibody. Since biotin is relatively small, several molecules can be conjugated to an antibody protein without altering its biological activity. The second step is to add a proprietary avidin–fluorophore conjugate. Avidin is a basic glycoprotein molecule with an extremely high affinity for biotin, so a strong link between the antibody and the fluorophore is rapidly formed.

Direct labelling involves mixing the fluorochrome and antibody together, under appropriate conditions; the fluorochrome becomes bound by reacting with NH_2 groups on the antibody protein (Goding, 1986).

An important consideration in comparing labelling methods is the intensity of the resulting fluorescence signal. With direct labelling it is not possible to bind more than three molecules of fluorescein (or many other dyes) directly on to an IgG antibody molecule, without diminishing its effectiveness. However, it is possible to attach five or six secondary antibodies to the same site. Thus, each antigenic site will have fifteen to eighteen dye molecules by indirect labelling, compared with a maximum of three dye molecules by the direct method. Likewise, the biotin–avidin

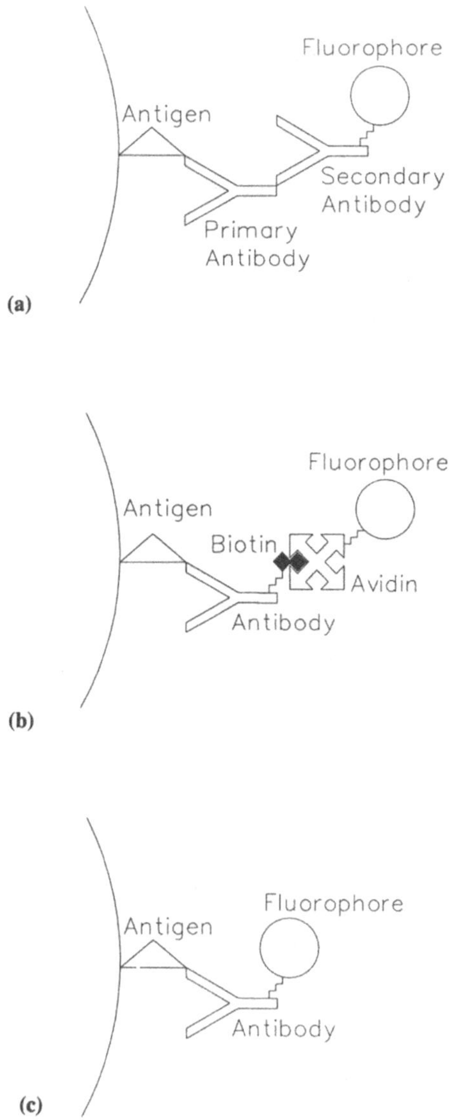

Figure 2.5 Methods of labelling antibodies fluorescently for the detection of specific cell surface markers (antigens): (a) the indirect, or sandwich, method; (b) the biotin–avidin bridge method; and (c) the direct conjugation method.

system also gives some signal enhancement, since a protein can react with several molecules of biotin, each of which in their turn can bind a molecule of avidin. Fluorescent labels derived from phycobiliproteins are a special case, as each phycobiliprotein molecule contains a large number of fluorophores of high absorbance and quantum efficiency. Therefore a directly conjugated antibody may have up to thirty associated fluorophores (i.e. it will be approximately ten times brighter than the equivalent fluorescein-conjugated antibody).

2.4.1.3 Viability labels. Demonstration of the presence or absence of an antigen may not, in itself, be sufficient to make a decision on product quality. It is also important to be able to differentiate between live and dead organisms, since in most cases dead bacteria are non-threatening; some measure of cell activity is required.

Viability tests by traditional microscopy are based on a cell's capacity to exclude acid dyes such as trypan blue, eosin, erythrosin, nigrosin or primulin, or certain basic dyes such as ethidium. The principle of the test is that these dyes do not normally cross intact cell membranes. Unfortunately, with some lethal treatments, cells may remain 'viable' by dye exclusion tests for several days. Whilst it is true that a cell which lets in the dye is dead, a cell which does not is not necessarily viable (Bhuyan *et al.*, 1976). This is especially true of organisms that have been killed by heat treatment, which is a situation commonly found in food processing.

An alternative approach to viability testing involves the nucleic acid stain acridine orange (AO), and exploits the presence of relatively high concentrations of RNA in actively growing cells. The AO concentration is adjusted so that when bound to nuclear DNA the fluorophore is exclusively in the intercalating, monomeric form, which exhibits green fluorescence. When bound to cytoplasmic RNA, AO forms a polymeric complex which exhibits red, metachromatic fluorescence. Thus, live cells should fluoresce red and dead cells green. Maintaining the correct staining conditions is critical, and again the method is unreliable with heat-treated organisms (O'Brien and McCully, 1981).

The lipophilic cationic dye, rhodamine 123, was first used as a stain for eukaryotic mitochondria. It is concentrated in mitochondria via the electrical potential across the membrane, which in turn is highly dependent on the energy metabolism process. A similar phenomenon is observed in bacterial membranes (Matsuyama, 1984) and is now being investigated as a viability indicator for flow cytometry (Kaprelyants and Kell, 1992).

Another effective way of obtaining viability counts is by using fluorogenic substrates to measure enzyme activity within the cell (Rotman and Papermaster, 1966). Examples of such reagents are esters of fluorescein, such as fluorescein diacetate (FDA). Commercial viability staining systems based on fluorescence derivatives have recently become available and give consistently reliable results with a wide range of microorganisms. The general principle of viability stains is that they readily cross cell membranes. Once inside a cell, they are cleaved by cytoplasmic esterases and liberate free fluorochrome. If the cell membrane is intact the fluorescent molecules are retained and concentrated in the internal matrix; if the cell membrane is damaged the fluorescence is lost. The intensity of the resulting intracellular cytoplasmic fluorescence therefore both differentiates between live and dead cells, and measures their relative metabolic activities. In other words, it provides information not simply on 'viability' but also on 'vitality' of a sample.

2.5 Research studies

2.5.1 *Identification of bacteria by light scatter and nucleic acid staining*

Using light scatter and a simple nucleic acid staining technique it is possible to distinguish between bacteria and other particles in a sample, and, to a limited extent, even to identify individual types of bacteria.

In order for a new technique in microbiology to gain recognition, it must be judged against the plate count method, which is still the most widely accepted yardstick. Detailed studies therefore have been undertaken comparing the accuracy of flow cytometry with conventional plate counting for measuring bacterial concentrations. Flow cytometry data from one such study (Pinder *et al.*, 1990) are shown in Figure 2.6. In this case, pure samples of *Pseudomonas fluorescens* were labelled with the nucleic acid stain ethidium bromide, after permeabilization with benzylkonium chloride. The figure shows two two-parameter scattergrams from two different

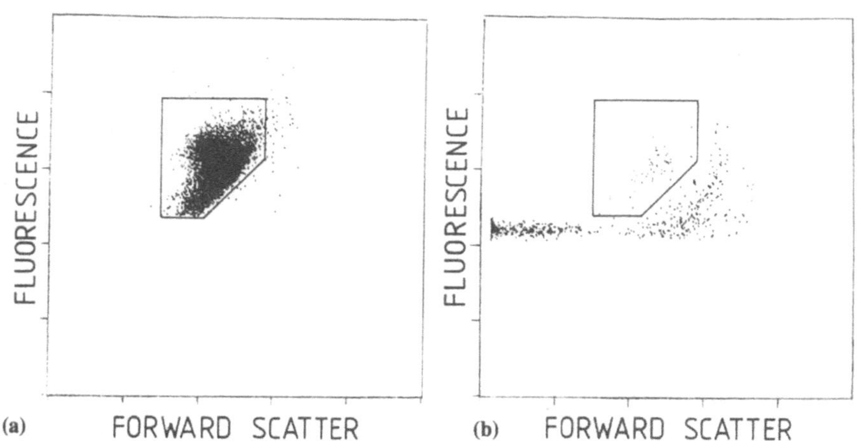

Figure 2.6 Scattergrams showing flow cytometric analysis of two samples of *Pseudomonas fluorescens* at concentrations of approximately (a) 5×10^6 per ml and (b) 1×10^3 per ml. The x-axis represents the intensity of light scattered in the forward direction, and the y-axis the intensity of fluorescent light emission; both axes are logarithmic, spanning five decades. Each black dot on the scattergram corresponds to an individual bacterium, and the density of dots represents the distribution over the sample. The solid line enclosing part of the scattergram was the computer 'window' used for analysis.

concentrations of the bacterium. In Figure 2.6(a), the concentration of the windowed region was $(4.9 \pm 0.3) \times 10^6$ ml^{-1}, whereas in Figure 2.6(b) it was $(1.2 \pm 0.1) \times 10^3$ ml^{-1}. The corresponding plate counts were subsequently determined as $(4.9 \pm 1.6) \times 10^6$ ml^{-1} and $(0.7 \pm 0.1) \times 10^3$ ml^{-1}, respectively. (In all cases, figures are quoted as mean $\pm$ standard error.)

Comparing Figure 2.6(b) with Figure 2.6(a), it will be noted that not only is the bacterial concentration much reduced, but so also is the proportion of computer-acquired events that fall within the window. This is because the time taken to acquire the data is much longer: spurious events, which occur with low frequency, now represent a significant fraction of the total number of events. These spurious signals may be due to non-specific contaminants and debris of finite size, small dye crystals or aggregates of material that take up dye to an intermediate extent. Nevertheless, even at a concentration as low as 10^3 ml^{-1} the events attributable to bacteria are clearly represented by a discrete population on the scattergram. Without specific labelling, such discrimination is not possible from an inspection of the single parameter histograms of scatter and fluorescence. This demonstrates the superiority of optical cytometry over other methods involving particle sizing or DNA fluorescence detection alone; it is the correlated multi-parametric nature of the measurement that makes such discrimination possible. The practical, lower concentration limit of flow cytometric detection from this study was $(2.2 \pm 0.8) \times 10^2$ ml^{-1}. The plate count corresponding to this sample was $(2.1 \pm 0.4) \times 10^2$ ml^{-1}.

Figure 2.7 shows the comparison of flow cytometric with plate count assay over five orders of magnitude of concentration, for four different types of bacteria: *Pseudomonas fluorescens* (gram-negative rods), *Aeromonas hydrophila* (gram-negative rods), *Streptococcus faecalis* (gram-positive cocci) and *Lactobacillus plantarum* (gram-positive rods). In all cases, the correlation coefficients are better than 0.99.

Data from stomached meat samples stained with ethidium bromide and analysed by flow cytometer are shown in Figure 2.8. In Figure 2.8(i) there are essentially two sub-populations. The region labelled 'C' corresponds to particulate matter and cell debris, which are inevitably present in such samples, span a broad range of particle sizes, and take up the ethidium bromide stain to some limited extent. However, a second region 'A' is clearly visible and distinct from 'C'. This sub-population corresponds to the intrinsic bacteria in the meat sample; the particles are small in size with a comparatively narrow distribution, but nevertheless highly fluorescent. There is good correlation between the computed cell concentration and the concentration obtained from an equivalent sample by traditional plate counting (approx. 1×10^4 cfu ml^{-1}).

The result shown in Figure 2.8(ii) illustrates the effect of adding a 'spike' culture of *Pseudomonas fluorescens* to the previous sample extracted from

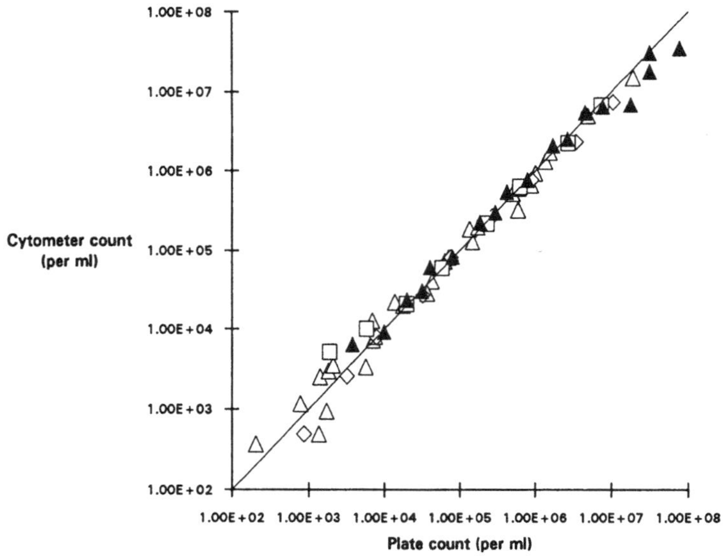

Figure 2.7 Comparison of the counts obtained by flow cytometry with that obtained by the conventional plate method for a range of concentrations of *Pseudomonas fluorescens* (△), *Aeromonas hydrophila* (▲), *Streptococcus faecalis* (◇) and *Lactobacillus plantarum* (□). The correlation coefficient of linear regression analysis was 0.992.

meat. A third region 'B' is now visible, which is very close to but just distinct from region 'A'. The ratio of the concentrations of extrinsic to intrinsic bacteria is 10:1, in good agreement with the level of spiking. Once again, it is only with correlated multi-parameter data that discrimination of these sub-populations becomes possible.

The above result suggests that it may be possible to differentiate between different types of bacteria, purely on the basis of their relative size and nucleic acid content. Figure 2.9 shows contoured two-parameter histograms of three bacterial species under similar flow conditions. Figure 2.9(a) is from a culture of *Pseudomonas fluorescens*. The contours indicate a comparatively modest variation of size and nucleic acid content; this bacterium is a rod typically 0.7 μm by 2 to 3 μm, and occurs mainly singly or in pairs. The distribution of *Streptococcus faecalis* (Figure 2.9(b)) shows a wider range, with a comet-like tail in the direction of increasing scatter and fluorescence. This is consistent with a tendency to form pairs or short chains. The peak of the distribution is at a marginally smaller size than Figure 2.9(a), which is also consistent with the fact that these cocci are ovoid and about 0.5–1.0 μm in dimension. By contrast, in the scattergram

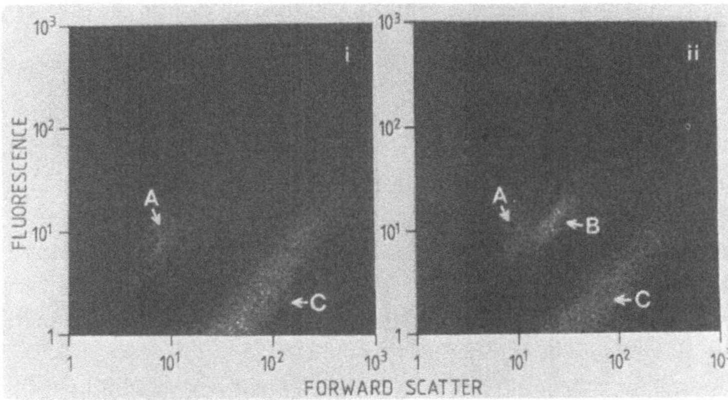

Figure 2.8 Contoured scattergrams from flow cytometer analysis of stomached meat samples stained with ethidium bromide. (i) Unadulterated sample showing population of intrinsic bacteria 'A', distinct from general debris 'C'; (ii) Sample with 'spike' of *Pseudomonas fluorescens*, showing a third region 'B' which is very close to but just distinct from region 'A'.

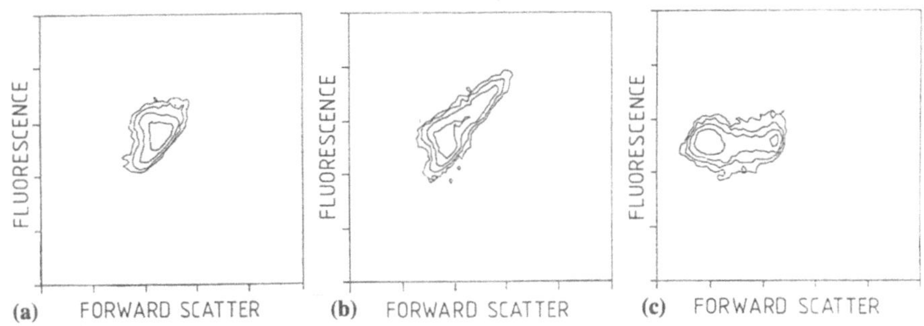

Figure 2.9 Comparison of flow cytometer analysis of cultures of (a) *Pseudomonas fluorescens*, (b) *Streptococcus faecalis* and (c) *Lactobacillus plantarum*, under similar conditions. From Pinder *et al.* (1990).

of *Lactobacillus plantarum* (Figure 2.9(c)), two apparently distinct subpopulations are observed. These bacteria are rods and the ratio of the lengths of the major and minor axes can be up to 10:1. The two principal size populations are therefore most probably associated with the two principal dimensions of this bacterium.

In this study, each bacterial species examined does, to some extent, exhibit its own characteristic 'signature' on the contour plot. The distributions are sufficiently distinct to enable an unknown pure culture to be identified from a selection of possibles, but not such as to enable mixtures of cultures to be resolved. It is well known that the morphology of bacteria can vary quite markedly, depending on culture conditions and whether they are in exponential or stationary growth phase. Therefore, although these results help to illustrate the resolving power of optical cytometry, they do not obviate the need for immunological protocols for the detection of specific types of bacteria.

2.5.2 *Identification of bacteria by immunofluorescent labelling*

Cytometric techniques based on fluorescently labelled antibodies are being developed in order to detect pathogens with species specificity. Antibodies, both polyclonal and monoclonal, have been raised against a number of food pathogens including *Salmonella* species, *Listeria monocytogenes*, *Campylobacter jejuni* and *Bacillus cereus* (Clarke *et al.*, 1992; Donnelly and Baigent, 1986; Tyndall *et al.*, 1985).

The following data are from part of a study using monoclonal antibodies that were selected for relatively species specificity to *Salmonella typhimurium* or *Salmonella enteritidis* (Clarke *et al.*, 1992). An antibody has also been produced which has a fairly broad specificity for a wide range of *Salmonella* serotypes, and virtually no cross-reaction with non-*Salmonella* species. The antibodies were fluorescently labelled with either fluorescein isothiocyanate (FITC) or R-phycoerythrin (R-PE) so that they could be used in combinations for two-colour detection by flow cytometry.

Figure 2.10 is a photomicrograph of a *S. typhimurium* culture labelled with FITC-conjugated antibodies (mercury epifluorescence illumination and × 100 objective). Since the antibodies attach to antigens on the surface, when the three-dimensional bacteria are viewed in two dimensions through the microscope the antibodies appear most concentrated around the periphery, hence the bright halo effect.

Flow cytometer analysis of a mixed *Salmonella* culture is presented in

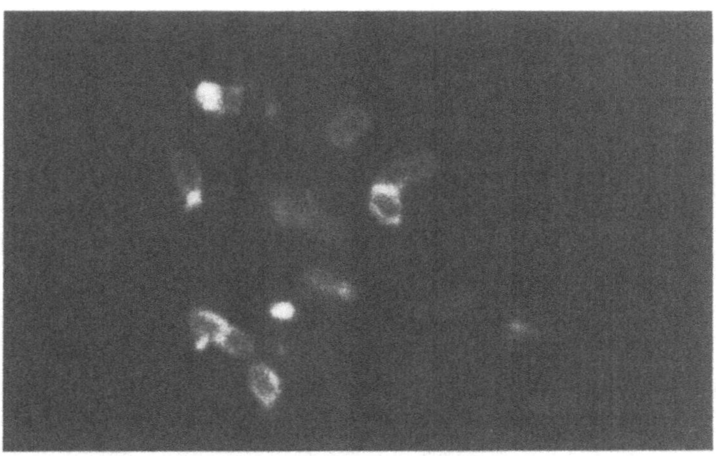

Figure 2.10 Photomicrograph of culture of *S. typhimurium* labelled with monoclonal antibodies conjugated to fluorescein isothiocyanate (FITC). Antibody fluorescence shows as a bright halo around each bacterium.

Figure 2.11. The sample was treated with FITC-labelled antibodies against *S. enteritidis* and R-PE-labelled antibodies against *S. typhimurium*. The figure is an isometric projection of a two-colour histogram, with FITC (green) fluorescence plotted along the *x*-axis and R-PE (red) fluorescence along the *y*-axis; the *z*-axis represents the cell abundancy. Three populations are visible ('A', 'B' and 'C'). 'A' exhibits strong green fluorescence, corresponding to the *S. enteritidis* population, whereas 'B' exhibits strong red fluorescence corresponding to the *S. typhimurium* population. A very small region 'C' is also visible, which corresponds to unlabelled cells and debris.

This analysis was repeated for a range of concentrations of *S. typhimurium* relative to a constant total concentration of all bacteria in the sample (Figure 2.12). For comparison, the expected count obtained by the plate method is also shown. It can be seen that the immunofluorescence/flow cytometry method gives reliable results for *S. typhimurium* down to below 10^4 ml^{-1}, even though this represents less than 0.1% of the total bacterial load.

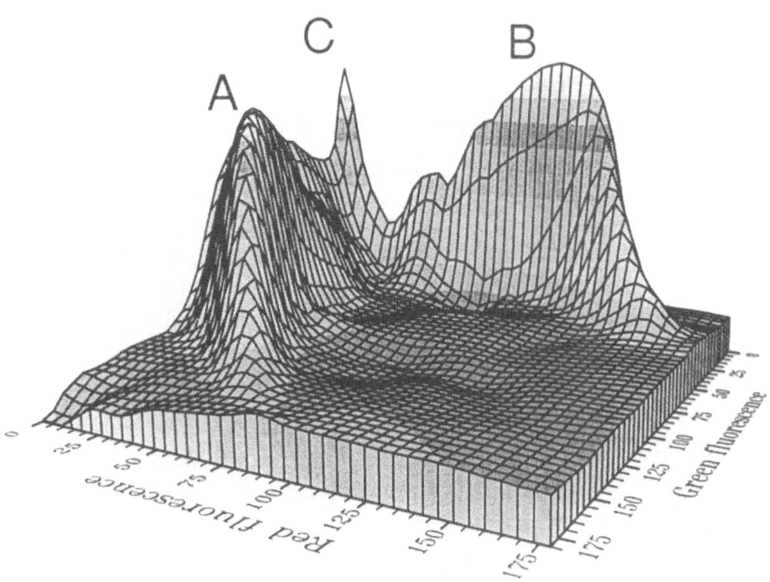

Figure 2.11 Scattergram showing the analysis of a mixed sample of bacteria by flow cytometry. The scattergram is plotted as an isometric projection, with the *z*-axis representing the relative number of cells. Population 'A' represents *S. enteritidis* cells that are stained green by FITC-labelled antibody, and 'B' represents *S. typhimurium* cells stained red by R-PE-labelled antibody. A small amount of unstained cells and other debris appear in population 'C'.

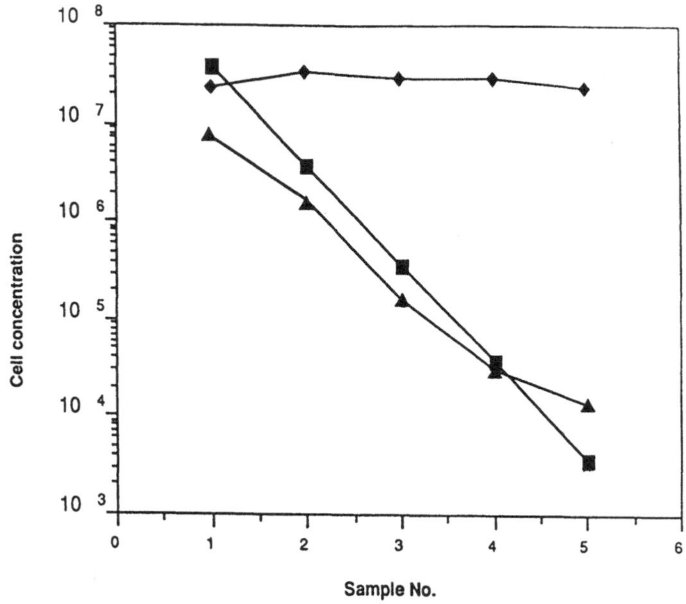

Figure 2.12 Graph showing the detection of *S. typhimurium* within a mixed *Salmonella* spp. population by flow cytometry. The total cell concentration remains constant (◆), while the relative *S. typhimurium* concentration is reduced ten-fold in each sample. The expected *S. typhimurium* concentration, as determined by the standard plate count method, is shown by (■) and the counts obtained by flow cytometry are shown by (▲).

2.6 Routine use of flow cytometry in quality assurance

The application of flow cytometry to the detection of microorganisms in routine industrial settings has required the development of simple, easy-to-operate instrumentation and ready-to-use reagents for a wide range of analyses. Since the mid-1980s developments in this area have intensified and there are now instruments and reagents commercially available which have been validated against standard culture techniques.

The following data have been generated in routine use of flow cytometry in industrial microbiology laboratories. In each case, the cytometer was a simple, mono-parametric instrument used in combination with viability substrate labels for counting viable microorganisms. Instrumentation and reagents for analysis of samples are commercially available in a ready-to-use form for each application cited.

2.6.1 Analysis of fruit preparations

Determination of yeasts and mould contamination is a key issue for producers and users of fruit products. Rapid detection of contamination of fruit products has been the subject of a number of recent studies with the

ChemFlow system (Chemunex S.A., Maisons Alfort, France; Brailsford and Gatley, 1992). The data in Table 2.3 show that this system is capable of detecting yeast in fruit products at levels of above 1 yeast per 10g of product in 24–48 hours. It is possible therefore to provide rapid positive release in fruit tanks prior to their use in yoghurt manufacture. This reduces the need for large cold store facilities for tanks awaiting positive release based on plate counts and improves the efficiency of fruit manufacture and dispatch.

2.6.2 *Analysis of milk products*

Table 2.4 shows ChemFlow detection of yeast in fermented milk products. Levels as low as 1 yeast per 10g of product were observed with the standard plate count method (Figure 2.13).

The cytometer data could also be correlated with the shelf life of products (Table 2.5). It can be seen that the cytometer cut-off levels correlated well with subsequent product quality. At a cytometer reading of less than 2000 counts per gram, product at end of shelf life was within specification. Above 2000 counts per gram all products were out of specification at the end of their shelf life. It should be noted that a similar correlation could not be achieved using the standard plate count, where counts of up to 5.2×10^5 cfu g^{-1} were obtained for both 'in-spec' and 'out-of-spec' products.

Through the determination of appropriate 'cut-off' levels it is possible therefore to use cytometer data to predict product quality at end of shelf life prior to shipment to the customer.

In addition to final product testing, the same system can be used as a process monitoring tool. Results presented in Table 2.6 show the levels of contamination compared to the time of detection. In both natural and fruit-based product types inoculation of 100–1000 yeasts per gram could be detected within three hours of product sampling. The slight differences

Table 2.3 Sensitivity of detection of yeast contamination in fruit products

Yeast inoculation	24 hour pre-incubation		48 hour pre-incubation	
	Plate result (+5 days)	Cytometer result	Plate result (+5 days)	Cytometer result
S. cerevisiae—0 cfu g^{-1}	–	–	–	–
S. cerevisiae—0.7 cfu g^{-1}	+	+	+	+
S. cerevisiae—6 cfu g^{-1}	+	+	+	+
S. cerevisiae—39 cfu g^{-1}	+	+	+	+
Z. bailii—0 cfu g^{-1}	–	–	–	–
Z. bailii—1 cfu g^{-1}	+	–	+	+
Z. bailii—9 cfu g^{-1}	+	+	+	+
Z. bailii—70 cfu g^{-1}	+	+	+	+

Table 2.4 Sensitivity of detection of yeast contamination in fromage frais by flow cytometry

Inoculation level (yeast/g)	Cytometer result (after 18 hours pre-incubation)
0	–
0.1	+
1	+
10	+
100	+

Table 2.5 Correlation between results by flow cytometer and standard plate count for shelf life of milk products

Cytometer count g^{-1} of product	Plate count cfu g^{-1} of product	Shelf life expiry date (~10°C)
100–2000	100–5200	Acceptable
> 2000	> 2500	Not acceptable

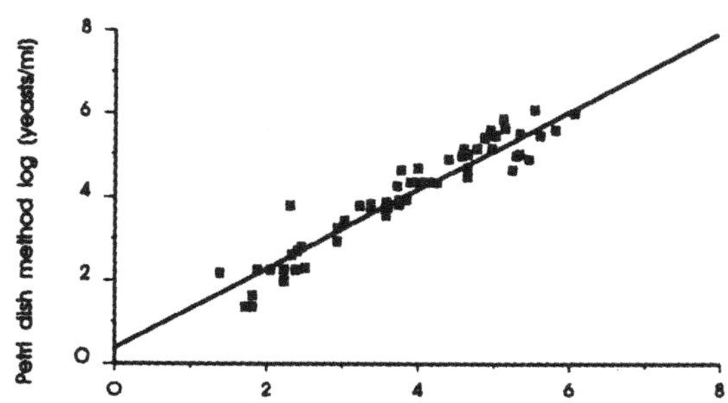

Figure 2.13 Comparison of the counts obtained by flow cytometry with that obtained by the conventional plate method for a range of concentrations of yeast in fermented milk products.

between two experiments reflects the yeast lag phase difference in the two product types. This approach can be used as a process line monitoring system or to allow rapid positive release of constituent products, e.g. white base prior to addition of fruit. Furthermore, it allows rapid intervention when a problem arises, thus minimizing plant down time and production losses.

2.6.3 *Starter culture analysis*

A key area in ensuring good product quality in fermented milk product manufacturing is the determination of starter culture activity, especially where a bulk starter may be used over several days. The method currently traditionally used for starter culture analysis is based upon the staining of viable organisms with methylene blue, followed by microscopic analysis.

Table 2.6 Incubation times required to reach a detectable level of yeast contamination in milk products

Inoculation level (yeast/g)	Product type	Incubation time required for detection (hour)
1	Natural	18
10	Natural	18
100	Natural	3
1000	Natural	2
1	Fruit based	18
10	Fruit based	6
100	Fruit based	2
1000	Fruit based	1

This method is both time consuming and subject to significant operator error. Using fluorescent viability staining coupled with optical cytometry, the metabolic activity of strains in the starter culture can be profiled in around 30 minutes.

Figure 2.14 shows a typical histogram obtained following analysis of lactic starter (*Streptococcus thermophilus* and *Lactobacillus bulgaricus*). The two cell populations are separated as a result of the different fluorescent signals given by cocci and bacilli. (In this case, fluorescent intensity is related to cell size as well as individual cell metabolic activity.) Computer analysis shows the separation of the two populations, and gives a percentage value for the different bacterial types making up the starter culture. Table 2.7 shows the correlation of the automated method with methylene blue microscopy.

2.6.4 *Shelf life prediction in salads and fruit juice manufacture*

In certain products, for example vegetable salads and fresh fruit, the growth of yeast contamination is a major factor in shelf life determination. Flow cytometry has been evaluated as a predictive tool to determine

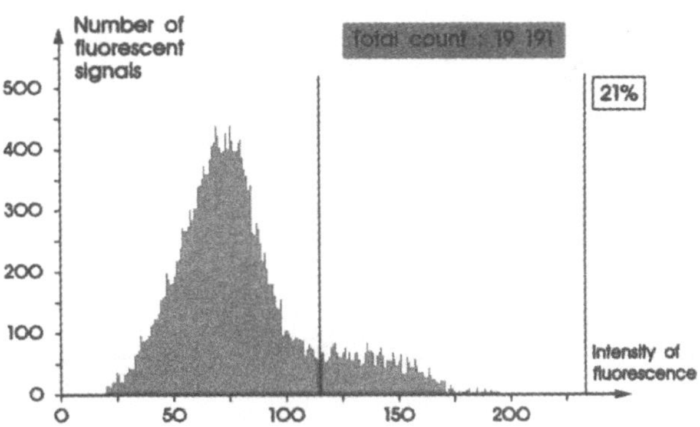

Figure 2.14 Fluorescence histogram of lactic starter culture, analysed by flow cytometer.

Table 2.7 Correlation between results by flow cytometer and standard methylene blue microscopy for lactic starter culture analysis

Day	Flow cytometer				Methylene blue
	Total flora ml^{-1}	*Lactobacillus* ml^{-1}	*Streptococcus* ml^{-1}	% *Lactobacillus*	% *Lactobacillus*
0	6.7×10^8	1.0×10^8	5.7×10^8	15	15–20
1	9.0×10^8	1.2×10^8	7.8×10^8	14	15–20
2	8.3×10^8	1.3×10^8	7.8×10^8	16	15–20
3	8.0×10^8	1.1×10^8	6.9×10^8	14	15–20

product shelf life. Samples from salad products were analysed after storage for seven hours at an elevated temperature of 30°C. Samples were also analysed after storage for 41 hours at 4°C, to determine the natural growth of yeast at normal storage temperatures. The results were used to calculate a spoilage index (count after incubation expressed as a factor of initial count). Determination of spoilage at elevated temperature showed good correlation with the spoilage of the product at normal storage temperatures, and provided good predictive data on product quality (Table 2.8).

Similarly, studies with fresh fruit juices (Figure 2.15) indicated that total yeast counts at filling were not necessarily predictive of product shelf life. High initial yeast counts did not necessarily correlate with subsequent product spoilage. The increase in counts following pre-incubation (i.e. the spoilage index) provided a more accurate indication of product quality.

2.7 Conclusions

Optical cytometric techniques are being used to develop robust, commercially viable techniques for routine use in microbiological monitoring.

Flow cytometry has been shown to be suitable for use throughout the manufacturing process, and provides rapid data to aid the management of fermentation processes, processing plant hygiene, raw material and final product quality. This allows the quality assurance manager to predict product quality throughout the production process, leading to reduced

Table 2.8 Flow cytometer analysis of yeast contamination in salad products—spoilage index determination

Product type	Spoilage index (7 hours at 28–30°C)	Spoilage index (41 hours at 4°C)
Coleslaw/natural	9.9	8.4
Coleslaw/fruit	0.8	1.0
Potato salad	0.5	0.5

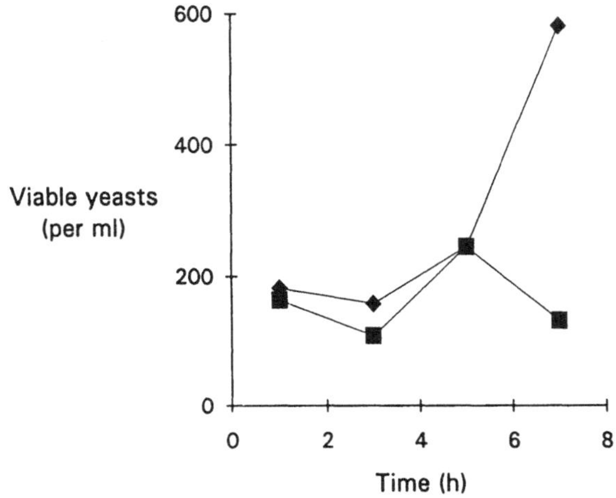

Figure 2.15 Data obtained by flow cytometry showing the growth (◆) or lack of growth (■) of yeast in fresh orange juice.

product recall, lower product losses and, ultimately, more cost effective operations.

For juice and salads manufacturers the commercial ChemFlow system can provide a rapid determination of total counts, and also the capability to generate more predictive data on the growth of yeast in the product and thus predict shelf life. This allows more careful control of production through raw material and process line testing and allows positive release of high quality product to the customer.

Further use of optical cytometric systems to detect microorganisms through the use of specific immunochemical labels, such as monoclonal antibodies, will ensure that this novel approach to rapid microbiological analysis provides new solutions for an even wider range of industrial users.

References

Bayer, E.A. and Wilchek, M. (1980) *Methods Biochem. Anal.*, **26**, 1–45.

Bhuyan, B.K., Loughman, B.E., Fraser, T.J. *et al.* (1976) Comparison of different methods of determining cell viability after exposure to cytotoxic compounds. *Exp. Cell Res.*, **97**, 275–282.

Brailsford, M. and Gatley, S. (1992) Rapid analysis of microorganisms using sensitive optical analysers. *Biotechnol. Internat.*, **34**, 186–192.

Clarke, R.G., Poulter, S.A.G., Langley, M., Pinder, A.C. and Lea, H.A. (1992) Detection of *Salmonellae* using fluorescent antibodies and flow cytometry. *J. Appl. Bacteriol.*, (in press).

Dean, P.N. (1990) Data processing. In *Flow Cytometry and Sorting* (eds Melamed, M.R., Lindmo, T. and Mendelsohn, M.L.). 2nd edn, Wiley-Liss, New York, pp.415–444.

Dean, P. and Pinkel, P. (1978) High resolution dual laser flow cytometry. *J. Histochem. Cytochem.*, **26**, 622–627.

Donnelly, C.W. and Baigent, G.J. (1986) Method for flow cytometric detection of *Listeria moncytogenes* in milk. *Appl. Environment. Microbiol.*, **52**, 689–695.
Festin, R., Bjorkland, A. and Totterman, T.M. (1990) Single laser flow cytometric detection of lymphocytes binding three antibodies labelled with fluorescein, phycoerythrin and a novel tandem fluorochrome conjugate. *J. Immunol. Methods*, **126**, 69–78.
Gatley, S. (1990) Digital microbiology—a radical approach to the design and development of a new rapid microbiology system. *Biotech. Forum Europe*, **6**, 478–482.
Gatley, S. (1991) Confocal optical scanning applied to the development of ultrasensitive microbiology analysers. *Biotech. Forum Europe*, **8**, 736–739.
Goding, J.W. (1986) *Monoclonal Antibodies: Principles and Practice*. 2nd edn, Academic Press, London.
Guilbault, G.G. (1989) Principles of fluorescence spectroscopy in the assay of food products. In *Fluorescence Analysis in Foods*. Longman Scientific and Technical, Harlow, Essex, pp.33–58.
Haugland, R.P. (1978) Covalent fluorescent probes. In *Excited States of Biopolymers* (ed. Steiner, R.F.). Plenum Press, New York, pp.29–58.
Hiebert, R.D. and Sweet, R.G. (1985) Electronics for flow cytometers and sorters. In *Flow Cytometry: Instrumentation and Data Analysis* (eds van Dilla, M.A., Dean, P.N., Laerum, O.D. and Melamed, M.R.). Academic Press, London, pp.129–162.
Kaprelyants, A.S. and Kell, D.B. (1992) Rapid assessment of bacterial viability and vitality by rhodamine 123 and flow cytometry. *J. Appl. Bacteriol.*, **72**, 410–422.
Kohler, G. and Milstein, C. (1975) Continuous culture of fused cells secreting antibodies of defined specificity. *Nature*, **256**, 495–497.
Kroll, R.G., Pinder, A.C., Purdy, P.W. and Rodrigues, U.M. (1989) A laser-light pulse counting method for automatic and sensitive counting of bacteria stained with acridine orange. *J. Appl. Bacteriol.*, **66**, 161–167.
Latt, S.A. and Langlois, R.G. (1990) Fluorescent probes of DNA microstructure and DNA synthesis. In *Flow Cytometry and Sorting* (eds Melamed, M.R., Lindmo, T. and Mendelsohn, M.L.). 2nd edn, Wiley-Liss, New York, pp.249–290.
Mackenzie, N.M. and Pinder, A.C. (1987) Flow cytometry and its applications in veterinary medicine. *Res. Veterinary Sci.*, **42**, 131–139.
Matsuyama, T. (1984) Staining of living bacteria with rhodamine 123. *FEMS Microbiology Letters*, **21**, 153–157.
Melamed, M.R., Lindmo, T. and Mendelsohn, M.L., (eds) (1990) *Flow Cytometry and Sorting*. 2nd edn, Wiley-Liss, New York.
Mullaney, P.F., van Dilla, M.A., Coulter, J.R. and Dean, P.N. (1969) Cell sizing: a light scattering photometer for rapid volume determination. *Rev. Sci. Instruments*, **40**, 1029–1032.
Munck, L., (ed.) (1989a) *Fluorescence Analysis in Foods*. Longman Scientific and Technical, Harlow, Essex.
Munck, L. (1989b) Practical experiences in the development of fluorescence analyses in an applied food research laboratory. In *Fluorescence Analysis in Foods* (ed. Munck, L.). Longman Scientific and Technical, Harlow, Essex, pp.1–32.
O'Brien, T.P. and McCully, M.E. (1981) *The Study of Plant Structure: Principles and Selected Methods*. Termacarphi, Melbourne.
Oi, V.T., Glazer, A.N. and Stryer, L. (1982) Fluorescent phycobiliprotein conjugates for analysis of cells and molecules. *J. Cell Biol.*, **93**, 981–986.
Olson, R.J., Frankel, S.L., Chisholm, S.W., *et al.* (1983) An inexpensive flow cytometer for the analysis of fluorescence signals in phytoplankton: chlorophyll and DNA distributions. *J. Exp. Marine Biol. Ecol.*, **68**, 129–144.
Patchett, R.A., Back, J.P., Pinder, A.C. and Kroll, R.G. (1991) Enumeration of bacteria in pure cultures and in foods using a commercial flow cytometer. *Food Microbiol.*, **8**, 119–125.
Pettipher, G.L. (1983) *The Direct Epifluorescent Filter Technique*. Research Studies Press, Letchworth.
Pinder, A.C., Purdy, P.W., Poulter, S.A.G. and Clark, D.C. (1990) Validation of flow cytometry for rapid enumeration of bacterial concentrations in pure cultures. *J. Appl. Bacteriol.*, **69**, 92–100.

Pinkel, D. and Stovel, R. (1985) Flow chambers and sample handling. In *Flow Cytometry: Instrumentation and Data Analysis* (eds van Dilla, M.A., Dean, P.N., Laerum, O.D. and Melamed, M.R.). Academic Press, London, pp.77–128.

Rodrigues, U.M. and Kroll, R.G. (1988) Rapid selective enumeration of bacteria in foods using a microcolony epifluorescence microscopy technique. *J. Appl. Bacteriol.*, **64**, 65–78.

Rost, F.W.D. (1992) *Fluorescence Microscopy*. Cambridge University Press, Cambridge.

Rotman, B. and Papermaster, B.W. (1966) Membrane properties of living cells as studied by enzymatic hydrolysis of fluorogenic esters. *Proc. Natl. Acad. Sci. U.S.A.*, **55**, 134–141.

Russ, W.D. (1990) Simultaneous, three colour immunofluorescence analysis by flow cytometry. *Focus*, **12**, 96–99.

Salzman, G.C., Crowell, J.M., Goad, C.A., Hansen, K.M., Hiebert, R.D., LaBauve, P.M., Martin, J.C., Ingram, M. and Mullaney, P.F. (1975) A flow-system multiangle light-scattering instrument for cell characterization. *Clinical Chem.*, **21**, 1297–1304.

Shapiro, H.M. (1988) *Practical Flow Cytometry*. 2nd edn, Alan R. Liss, New York.

Steen, H.B., Boye, E., Skarstad, K., Bloom, B., Godal, T. and Mustafa, S. (1982) Applications of flow cytometry on bacteria: cell cycle kinetics, drug effects, and quantitation of antibody binding. *Cytometry*, **2**, 249–257.

Thorell, B. (1981) Flow cytometric analysis of cellular endogenous fluorescence simultaneously with emission from exogenous fluorochromes, light scatter, and absorption. *Cytometry*, **2**, 39–43.

Tyndall, R.L., Hand, R.E., Mann, R.C., Evans, C. and Jernigan, R. (1985) Application of flow cytometry to detection and characterisation of *Legionella* spp. *Appl. Environment. Microbiol.*, **49**, 852–859.

van Dilla, M.A., Langlois, R.G. and Pinkel, D. (1983) Bacterial characterisation by flow cytometry. *Science*, **220**, 620–621.

Waggoner, A.S. (1990) Fluorescent probes for cytometry. In *Flow Cytometry and Sorting* (eds Melamed, M.R., Lindmo, T. and Mendelsohn, M.L.). 2nd edn, Wiley-Liss, New York, pp.249–290.

Wheeless, L.L. and Kay, D.B. (1985) Optics, light sources, filters and optical systems. In *Flow Cytometry: Instrumentation and Data Analysis* (eds van Dilla, M.A., Dean, P.N., Laerum, O.D. and Melamed, M.R.). Academic Press, London, pp.21–76.

Wilson, T. (1990) *Confocal Microscopy*, Academic Press, London.

Wilson, T. and Sheppard, C.J.R. (1984) *Theory and Practice of Scanning Optical Microscopy*. Academic Press, London.

3 Infrared spectroscopic methods

R.H. WILSON and E.K. KEMSLEY

3.1 Introduction

There is a growing interest in industry in process control systems employing infrared methods. In the food sector, this interest has primarily been associated with near infrared (NIR) systems which have gradually moved from the laboratory environment to the factory floor. Mid infrared (MIR) methods are now becoming increasingly used and in the future may become extremely important as process control devices. The object of this chapter is to describe the current status of NIR and MIR process control technology, with special emphasis on developments that have affected the MIR region particularly. The future potential of infrared methods will also be considered.

3.2 Infrared spectroscopy

Infrared spectroscopy is a form of vibrational spectroscopy (see, for example, George and McIntyre, 1987). An infrared spectrum arises from the interaction of infrared radiation with the molecular bonds within a sample. The spectrum of any material will show absorptions at certain wavelengths that are characteristic of the chemical entities present. These absorptions can be used qualitatively to show the presence of a particular analyte, or quantitatively to determine the concentration of the analyte(s) in the sample.

In simple terms, the bond joining two atoms (masses m_1 and m_2) can be thought of as a tiny 'spring'. This spring has a natural 'stiffness', that is, a resistance to extension denoted by the force constant, k, which is related to the strength of the chemical bond. Given sufficient energy, this system can be set into vibration at a frequency, f_0, which is predicted from classical mechanics:

$$f_0 = \frac{1}{2\pi} \sqrt{\frac{k}{\mu}} \qquad (3.1)$$

where μ is the reduced mass, given by:

$$\mu = \frac{m_1 m_2}{m_1 + m_2} \tag{3.2}$$

The energy of such oscillations is quantized to satisfy:

$$E_n = (n + ½)\, hf_0 \tag{3.3}$$

where the vibrational quantum number $n = 0, 1, 2, \ldots \infty$, h is Planck's constant, and f_0 is the classical vibration frequency from equation (3.1) above.

At room temperature most molecules exist in the vibrational ground state, E_0. Thus,

$$E_0 = \frac{1}{2}\frac{h}{2\pi}\sqrt{\frac{k}{\mu}} \tag{3.4}$$

If given sufficient energy, a transition to the first excited vibrational state can occur. The energy of the first excited state, E_1, is given by

$$E_1 = \frac{3}{2}\frac{h}{2\pi}\sqrt{\frac{k}{\mu}} \tag{3.5}$$

The energy ΔE absorbed during the transition is therefore:

$$\Delta E = E_1 - E_0 = \frac{h}{2\pi}\sqrt{\frac{k}{\mu}} \tag{3.6}$$

Infrared radiation of frequency ν and energy $h\nu$ can supply the energy required for a transition, provided that:

$$\nu = \frac{h}{2\pi}\sqrt{\frac{k}{M}} \tag{3.7}$$

In infrared spectroscopy, the frequency ν is usually expressed in wavenumbers (cm^{-1}). The frequency in wavenumbers, σ, can be obtained by dividing equation (3.7) by the velocity of light, c, in cm s^{-1}.

$$\sigma = \frac{1}{2\pi c}\sqrt{\frac{k}{\mu}} \tag{3.8}$$

Spectral collection involves irradiation of a sample with 'broadband' infrared radiation, comprising a range of frequencies. Absorption occurs only at wavenumbers satisfying equation (3.8). In practice infrared absorptions are not discrete lines but tend to be broadened. This is the result of simultaneous rotational transitions (George and McIntyre, 1987).

The usefulness of infrared spectroscopy as an analytical tool lies in its ability to determine molecular structure and composition. Equation (3.7) shows that the absorption frequency for a given bond depends upon its strength and the masses of the atoms forming the bond. For example, the absorption of C = O occurs at a different frequency to C – O; despite identical reduced masses for these bonds, C = O has a higher force constant and thus the absorption frequency is higher. A C – H bond has a much lower reduced mass, and its absorption occurs at an even higher frequency. Infrared spectroscopy can thus be used to determine which functional groups are present in a sample.

Most foods are made up of carbohydrates, proteins, fats and water. Each of these has a distinct and characteristic functional group (Table 3.1) so that each gives rise to an infrared absorption indicative of the presence of that component in the sample. Figure 3.1 shows a typical infrared absorbance spectrum, obtained from a sample of wheat flour. Furthermore, infrared spectra are quantitative. The intensity of an absorption is directly proportional to the amount of absorbing species, making infrared spectroscopy attractive as an analytical method in the food sector. However, the widespread use of the technique for food analysis has been hindered in the past by technical limitations which have meant that the most useful spectroscopic region, the MIR, where the distinct fundamental vibrations occur, could not be used. Instead much effort has gone into the NIR region where sampling is more straightforward. Absorptions in the NIR arise from overtones (multi-quanta transitions) and combinations of the fundamental vibrations. Strictly forbidden transitions, these absorptions are weaker and really only occur for bonds involving an atom coupled to hydrogen. The reduced information content and potential complexity has meant that whereas sampling is easy, spectral interpretation is not, and

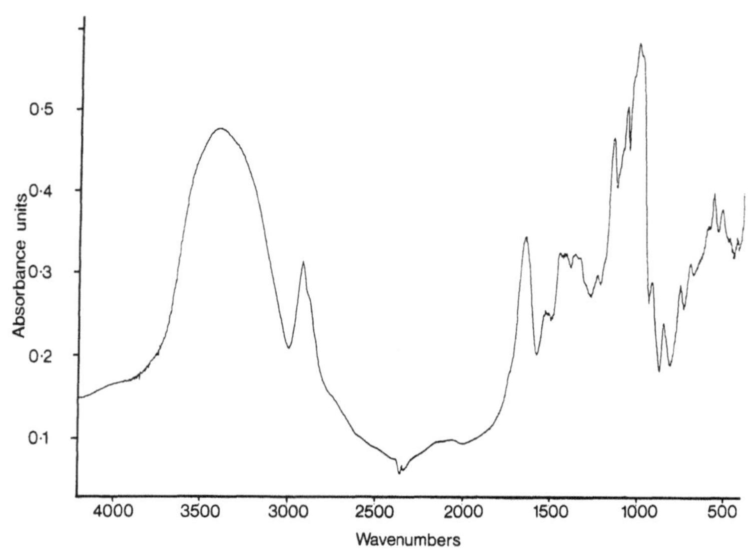

Figure 3.1 Typical infrared absorbance spectrum, obtained from a sample of wheat flour.

Table 3.1 Characteristic functional groups of food components

Wavenumber of band	Chemical functionality	Food component
3600–3200	O–H stretch	Water, carbohydrates
3030	C–H of *cis* double bond	Unsaturated fat
3000–2700	C–H	Fats, proteins, carbohydrates
1745	C=O, ester	Fats
1725	C=O, ester	Pectin
1600–1700	C=O, acidic	Fatty acids, acetic acid
1640	O–H (bend)	Water
1650	C=O, amide I	Protein
1550	N–H, amide II	
1400–900 (complex)	C–O, C–C	Carbohydrates (fingerprints), fats

the NIR is heavily dependant on applied statistics. In recent years, however, there have been important developments that now mean the fundamental or MIR region can be exploited for analysis (Wilson, 1990). These developments have also led to increased interest in the use of MIR for on-line analysis and process control. These advances include Fourier transform (FT) methods and internal reflectance sampling, which will be discussed in the following sections.

3.2.1 *Infrared spectrometers*

In a dispersive spectrometer (Figure 3.2), the infrared radiation from the source is divided into two by a suitably shaped mirror. One beam passes through the sample, whilst the other traverses a reference channel containing a matched compensating cell. The two beams then pass via a chopper into the monochromator, where each is split into its component wavelengths by a prism or grating. A narrow range of wavelengths is allowed to exit the system by the use of a slit of appropriate width. At the detector the two signals are ratioed. With such a system an entire spectrum can be generated by rotating the prism or grating, thus allowing the absorption at each wavelength to be determined.

The output from a dispersive spectrometer is the ratio of the light intensity of the reference channel (I_0) to that of the sample (I). This ratio is called the transmittance and is usually defined as:

$$\%T = \frac{I}{I_o} \times 100 \tag{3.9}$$

I_0 is formally the incident light intensity at the sample, but compensation is usually made for energy loss due to properties of the cell and solvent absorption, by use of the reference channel. I is less than I_0 when the sample absorbs at some wavelength, λ. If the sample cell contains an absorber of concentration, c, in a cell of thickness (optical path length) l:

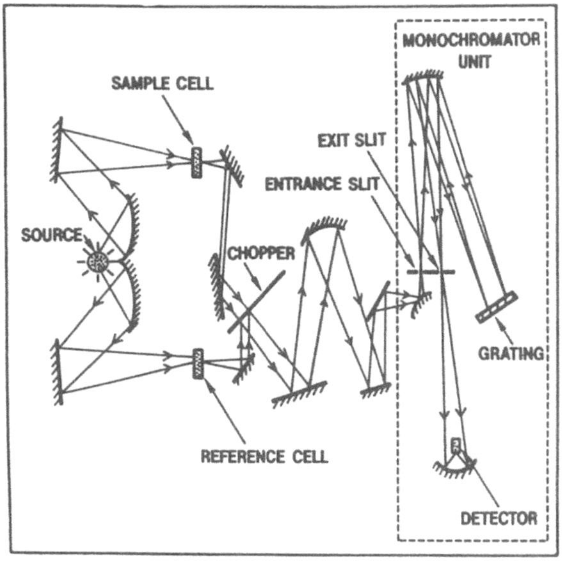

Figure 3.2 Schematic of a dispersive infrared spectrometer.

$$I = I_0\, 10^{-\varepsilon c\, l} \tag{3.10}$$

where ε is a physical constant of the absorber called the absorptivity. Rearranging and taking logs leads to:

$$A = \log_{10} \frac{I}{I_0} = \varepsilon\, c\, l \tag{3.11}$$

where A is the absorbance (expressed in absorbance units, AU). This equation is the Beer–Lambert relationship which relates absorbance linearly to concentration, and provides the basis of quantitative infrared spectroscopy.

Although dispersive instruments have their advantages and indeed are widely used in the NIR, for the MIR, Fourier transform methods are a necessity (Griffiths, 1975). A conventional scanning monochromator at any given time transmits only a small amount of the available radiation which is passed by the instruments slits. To collect a complete spectrum also takes a significant time as the dispersing element has to be mechanically moved. In a Fourier transform instrument (Figure 3.3), radiation from the source is directed to a beam splitter (typically a coating of germanium upon a salt substrate) where the incident light is divided. One beam is transmitted to a fixed mirror and the other reflected to a mirror mounted on a moving bearing. After reflection at each mirror the light recombines at the beam splitter, some passes to the sample position, and

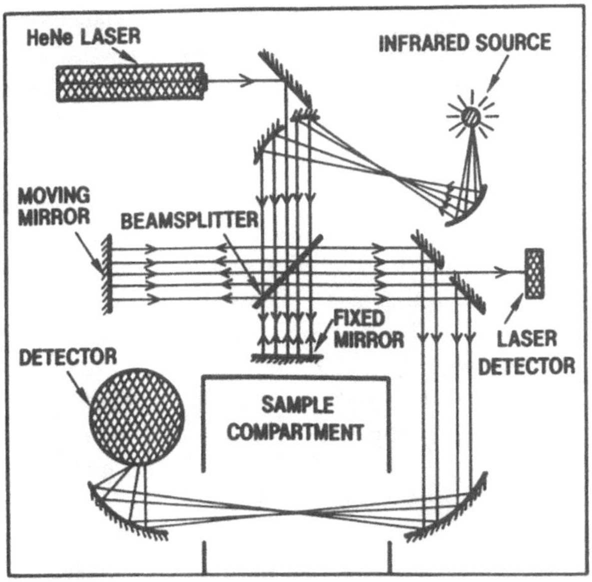

Figure 3.3 Schematic of an FT–IR spectrometer.

hence to the detector. An optical path difference between the two beams can be induced by making the distance from the beam splitter to the two mirrors unequal by moving the movable mirror. Light of any given wavelength, λ, returning to the beam splitter after reflection will be in-phase and will interfere constructively when the path difference is an integral multiple of λ. If however, the mirror is moved such that the path difference is a half integral multiple of λ, then at the beam splitter the returning light will be out-of-phase and destructive interference will occur. In summary, constructive interference occurs when the path difference is $n\lambda$ and destructive at $(n + 0.5\ \lambda)$, where $n = 0, 1, 2. \ldots$ As the mirror is moved, intensity maxima and minima at any given wavelength are seen in the beam emerging from the interferometer. This sinusoidal intensity wave is called an 'interferogram' and the device producing it is the Michelson interferometer. For a real infrared source, which produces a 'broadband' of many infrared wavelengths, the interferogram comprises the sum of all the individual interferograms of the many source wavelengths. It exhibits a large maximum intensity at zero path difference, but its intensity rapidly decays with path difference (also called optical retardation) and 'beats' may be observed. In order to generate a spectrum (intensity versus wavelength) the interferogram (intensity versus retardation) is subjected to the mathematical process of Fourier transformation. This is not strictly necessary in that the interferogram contains all the same information as the spectrum and could be used directly. However, interferograms are not easily interpreted by eye and so the Fourier transform is used.

In an FT-IR spectrometer, the entire spectrum is collected in the time it takes to move the mirror to its maximum travel, which may be less than a second. Therefore, in the time taken by conventional dispersive instrumentation to collect a single spectrum, the interferometer can collect many, typically 60–100 min^{-1}. These are co-added to provide an improvement in signal-to-noise in the final spectrum of $\sqrt{n}$, where n is the number of co-added spectra. This process is known as signal averaging. In addition, the absence of slits in the system allows a much greater fraction of the source energy to reach the sample. These advantages combine to allow spectra of good quality to be obtained from samples and sampling accessories that are poorly transmitting or have a low energy throughput. A further feature of FT-IR spectroscopy is the high reproducibility in the wavelength scale resulting from the use of an He–Ne laser reference. The laser radiation, following an identical path to the infrared through the interferometer, is used to track the position of the moving mirror and to trigger the A-D converter. A full description of interferometry can be found in a number of extensive works in the literature (see, for example, Griffiths, 1975). In an FT-IR spectrometer single beam operation is employed. The values of I_0 at every wavelength are collected by measuring a spectrum of the sample cell without the sample present. This 'background' is stored digitally and all sample spectra are ratioed to it.

3.2.2 Approaches to sampling

Whatever the spectrometer system or wavelength range, at some point there must be an interface that allows interaction between the infrared and the sample. Transmission methods are probably the most widely used in spectroscopy generally (Miller and Stace, 1979). The simplest transmission cells consist of two transparent windows held at a fixed separation. The sample is placed between the windows for analysis. For the NIR, the cell path length can range from about 0.5 mm to a few cm, which makes cell design relatively easy. For on-line use, fluid can be flowed through the cell. For solid sampling diffuse reflectance is employed in the NIR with great effect.

Such methods are of little use in the MIR, particularly for on-line sampling. In diffuse reflectance, the wavelengths of light in the MIR tend to be of a similar order of magnitude to the particle size of many powdered samples, so that scattering effects become more pronounced. The presence of anomalous dispersion in the MIR can also lead to unusual effects. The overall result is that MIR diffuse reflectance spectra can be difficult to interpret and quantify. Furthermore, most commercial diffuse reflectance accessories are designed for small, homogeneous samples and cannot be used on-line.

Other alternatives available for laboratory sampling such as photoacoustics (Belton and Wilson, 1990) and pellet methods are also not suitable for on-line situations. Therefore, solid sampling in the MIR is difficult and represents a real hurdle to on-line implementation of MIR spectroscopy. Proposed solutions to this problem include emission, stimulated emission and other photothermal methods (Jones and McClelland, 1990a,b), as well as modified diffuse reflectance accessories that are more suitable for process applications.

Transmission methods are also problematic in the MIR. Because the absorptivity of samples tends to be much greater, short path lengths of less than 10 μm must be employed, making cells difficult to build, fill and clean. Transmission methods are also inappropriate for highly scattering or opaque samples, or the wide range of semi-solid materials which cannot be sampled by diffuse reflectance (Belton and Wilson, 1990).

Internal reflectance methods (Harrick, 1987) have, therefore, found favour in the MIR where the technique is called attenuated total reflectance (ATR). In ATR (Figure 3.4) the infrared radiation is directed into a crystal of high refractive index (n_1) at such an angle that it arrives at the interface of the crystal with the surrounding medium (refractive index n_2) at an angle (θ_i) greater than the so-called critical angle (θ_c).

$$\theta_c = \mathrm{Sin}^{-1} \frac{n_2}{n_1} \tag{3.12}$$

This is the condition for total internal reflection. The light is reflected at the interface and is trapped inside the crystal whose geometry is designed to ensure that all rays are internally reflected and emerge from the crystal after 6–10 reflections. At every internal reflection site, an evanescent field penetrates into the surrounding medium. If the medium is absorbing, then

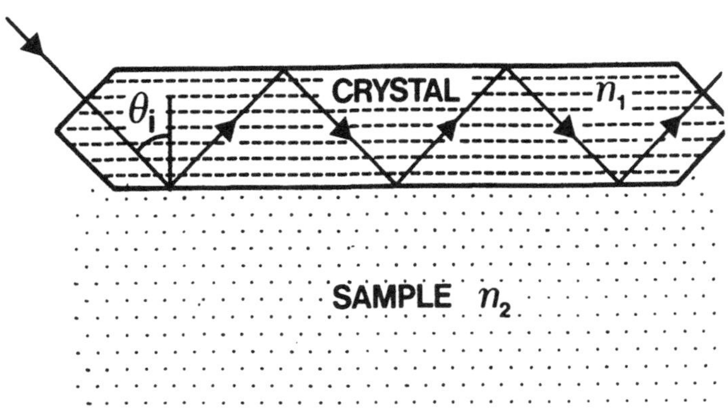

Figure 3.4 The ATR sampling technique.

there is energy transfer to the field and the infrared is attenuated. The penetration is small (a few μm) and depends on the ratio n_2/n_1 and θ_i. Interaction no longer involves transmission in the normal sense and consequently spectra can be obtained from samples that are opaque and highly scattering; they must only make good optical contact with the crystal. The path length for such devices overall is very short and reproducible, thus overcoming the major limitations with transmission methods. ATR cells exist in a number of shapes and formats. Some are suited to rapid off-line sampling whilst others are designed to be used on-line in a flow-through mode.

3.2.3 Quantification

In certain cases, each component of interest within a sample may give rise to a well-resolved, characteristic absorption. In these circumstances, simple calibrations based on the Beer–Lambert relationship (equation (3.11)) can be used for quantitative analysis. More often, however, absorbance broadening means that spectral features are severely overlapped; that is, the absorbance at any wavenumber does not depend on the concentration of a single analyte, but on the sum of contributions from several different analytes. Calibration becomes more complicated, requiring sophisticated multivariate methods and larger calibration sets. The weak and broadened absorbances found in the NIR exacerbate the problem, and quantitative work relies heavily on statistical procedures. MIR spectra are better resolved, and it is often possible to identify individual absorbances, even in mixtures of analytes of similar chemical structure (for example, sugar mixtures). Nevertheless, band overlap also occurs, and multivariate methods are frequently required.

Straightforward multiple linear regression methods (the so-called K- and P-matrix approaches) have been used extensively, and can be quite successful. Both techniques seek to solve an equation relating the matrix of absorbance data (measured at a number of selected wavenumbers) to the matrix of concentration data of the calibration set (Brown *et al.*, 1982). The matrix equations in each approach differ slightly; a consequence of this difference is that the P-matrix method is better able to cope with impurities or unexpected components. 'Over-determination' is common practice in both approaches: that is, substantially more calibration spectra or absorbance measurements than the required minimum are used, in order to improve the robustness of the calibration.

Two more complex methods have also received attention: principal components regression (PCR) and partial least squares (PLS) (see for example, Martens and Naes, 1989). In contrast to the K- and P-matrix methods, which generally use a limited number of absorbance measurements, the 'data compression' methods of PCR and PLS can be applied to whole spectra. In both approaches, 'factors' are determined, which represent uncorrelated sources of variation in the calibration set. The factor 'loadings', which resemble spectra, show the correlations existing between data points. Linear combinations of the factor loadings can be used to reconstruct the original calibration spectra; the multiplicative values required for reconstruction are termed factor 'scores'. In both PCR and PLS, the matrix of scores essentially forms a new data set, on which the concentration data are regressed to obtain a calibration. However, in PLS the concentration data are additionally used in the determination of the loadings, to take into account not only spectral but also concentration values. PLS is likely to give superior results for trace analysis; in PCR, major constituent effects can dominate.

Spectrometer manufacturers usually offer a range of quantitative analysis packages to accompany their instruments. Some of these can be adapted for process monitoring, although this is not yet standard. There are also software companies which specialize in statistics packages for spectroscopy. For example, the 'Unscrambler' package, available from CAMO of Trondheim, Norway, is widely used in NIR spectroscopy for PCA, PLS and exploratory data analysis.

3.3 Instrumentation for process monitoring and control

Dispersive and FT-IR instruments have both been used in process situations to a greater or lesser extent. The low throughput, slow acquisition times and adverse beam geometry already discussed limit the usefulness of dispersive spectrometers as process control devices in the MIR. However, some NIR instruments equipped with optical fibres have been used for on-line monitoring, although these are likely to face increasing competition from the recent development of fibre-optic FT-IR. Indeed continuing improvements in several areas of FT-IR instrumentation appear certain to increase its importance in the future. Selective-wavelength devices, based on narrow bandpass filters, are conceptually and mechanically more simple than dispersive or FT-IR instruments. Although not spectrometers in the true sense, these are well established in the process field, and will also be considered.

3.3.1 Selective-wavelength instruments

Selective-wavelength devices represent the most widely used infrared instrumentation currently found in the process field. Both NIR and MIR instruments are available, for a wide variety of applications (Hyvärinen and Niëmela, 1990). Although sampling methods may differ to suit the spectral range and the type of sample, the basic principles of operation are the same. Figure 3.5 shows a typical NIR instrument. An infrared source (for example a hot wire filament) provides broadband radiation in the appropriate spectral region. After sampling, narrow bandpass infrared filters are used to select a limited number of wavelengths at which absorbance measurements are made. Detectors are typically pyroelectric (MIR) or photodiode (NIR). The absorbance measurements are transformed to concentration data, using a calibration stored in the device's central processor unit.

The number of filters and their passbands are chosen to suit the particular application for which the instrument is designed. For some applications, where the concentration of one component only is to be determined, a single filter may suffice. More often, however, a reference wavelength is needed, or a multi-component system is of interest. Consequently, data from several wavelengths must be collected; multiple filters with different passbands are required. In such cases, the filters may be conveniently mounted on a chopper, so that each wavelength is presented successively to the detector. The resulting signal is electronically decoded, and passed to the central processor for presentation in the desired form. Alternatively, the infrared beam may be split after sampling, and each portion diverted to an individual detector fitted with a filter in the detector window. A chopper modulates the infrared beam before it is split. Thus, the detector signals are always synchronized, and are passed in parallel to the central processor. By replacing the large-aperture chopper-mounted

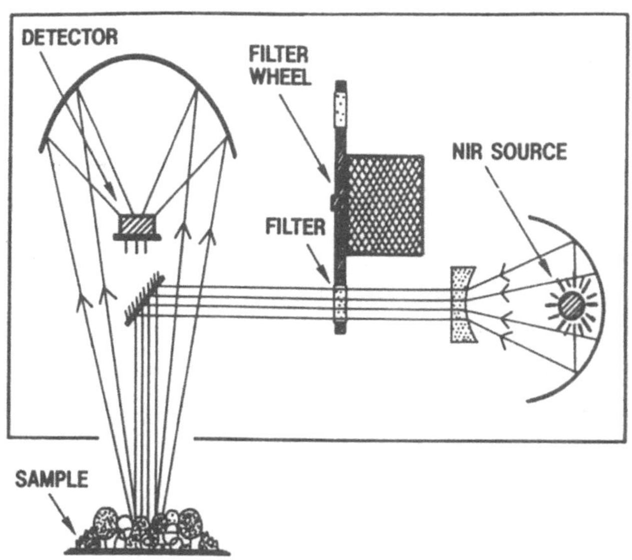

Figure 3.5 Schematic of a typical NIR selective-wavelength instrument.

filters with small detector window filters, this configuration offers significant cost benefits.

NIR selective wavelength devices are well established in the process field. Instruments have been developed for the analysis of a wide range of sample morphologies, from powdered, crushed and extruded solids, to liquids and emulsions. Transmission and reflectance techniques are generally used for sampling. Devices are available for the quantitative determination of components such as moisture, fat, oil and protein in products including instant coffee, skim milk powder, breakfast cereals, potato crisps and flour. For example, the MM55 Biscuit Gauge from Infrared Engineering (Maldon, Essex) is a NIR multi-wavelength instrument, adapted for the measurement of moisture in the biscuit industry. Mounted approximately 3 cm above the conveyor, the Biscuit Gauge uses a non-contact reflectance sampling method, therefore no risk of contamination is introduced. Also incorporated is a system for detecting the leading and trailing edges of the biscuits, so that data are only collected and processed when a biscuit is present in the sampling region. Moisture contents in the range 0–80% can be determined with an accuracy of ± 0.1%. Also available are complete process control packages. The Flour Control System from Infrared Engineering uses the basic MM55 sensor to view the sifted flour stream. Moisture and protein measurements, accurate to ± 0.15%, are made several times a second. The central control unit processes the data, and provides continuous adjustment to the feeder from the gluten reservoir, thus maintaining the protein level to specification.

MIR devices are less widespread, due to the practical problems presented by operating in the MIR region. Detectors are less sensitive, throughput is generally lower, and therefore signal-to-noise ratios are poorer than can be obtained in the NIR. Moreover, sampling is much more difficult. Transmission techniques are effectively ruled out for process applications, as passing a typical process stream through the short cell path lengths required in the MIR is impractical. Diffuse reflectance methods may be possible, and some research is currently underway to develop the technique for process use, particularly in conjunction with FT-IR, as reflectance methods represent the most promising approach for the on-line sampling of solids. MIR selective-wavelength instruments available at present invariably use ATR as the sampling technique, and are restricted to analysing liquids. Typical applications include the measurement of acetic acid in vinegar, sugar-to-acids ratios in fruit juice, and carbon dioxide in soft drinks. For example, the LAN-II Process Monitor manufactured by General Analysis Corporation (South Norwalk, Connecticut, USA) measures the sugar and acids contents of fruit juice, with accuracies of ± 0.2% for the sugar, and ± 2% for the acids. Three filters are employed: a 2940 cm^{-1} (3.4 μ) filter to measure the sugar, a 2560 cm^{-1} (3.9 μ) filter for acids, and a 4000 cm^{-1} (2.5μ) reference filter. A cylindrical sapphire ATR element is used for sampling. The LAN-II meets environmental and safety conditions and can be plumbed directly into the fruit-juice process stream.

Selective-wavelength devices have successfully entered the process industry. Ruggedization presents few problems: the optical components can be mounted inside robust, sealed modules as little as 30 × 20 × 15 cm in size, with similar modules housing the associated hardware, such as power supplies, controls and displays. However, there are limitations to these instruments. Most devices are application specific: the number of filters and their passbands are carefully chosen to suit the particular process for which the instrument is designed. Furthermore, the spectral information obtained by selective-wavelength devices is restricted. In contrast, a spectrometer collects and processes data from a wide wavelength range; it is not limited to a single application, and is more able to handle unexpected results or identify contaminants, which may be beyond the scope of the filter instrument.

3.3.2 FT-IR instruments

Fourier transform instruments have yet to become established in the process field. This is partly due to resistance to change encountered in the process industries, but there are also very real shortcomings in the instruments themselves, which are generally designed for use in the laboratory and are unable to meet the increased demands of the process environment. However, manufacturers of laboratory spectrometers recognize that this area represents a potentially lucrative market, and considerable effort is being expended to improve both the ruggedness and reliability of their instruments. Several different approaches can be taken to on-line FT-IR. These include flow-through sampling systems, beam-steering remote sampling systems, and fibre-optics.

A number of manufacturers now produce flow-through cells intended for process applications, based on modifications of conventional sampling accessories. This type of accessory is mounted in the spectrometer sample compartment. Liquid is drawn from the process line and pumped through the cell for sampling. The most usual sampling method is ATR. Cells are available in a variety of geometries; crystal shapes include cylindrical, square cross-sectional and rectangular cross-sectional. A typical flow-through ATR cell employing a cylindrical crystal is shown in Figure 3.6. Crystal materials are generally chosen from Zn, Se, Ge and Si. Unfortunately, for food applications, safety considerations may prevent such a choice. Sapphire represents a possible alternative, as it is non-toxic, chemically inert and extremely strong. Its use as an ATR element has already been demonstrated in selective-wavelength devices. Its optical properties, however, are not ideal, and cells employing sapphire crystals

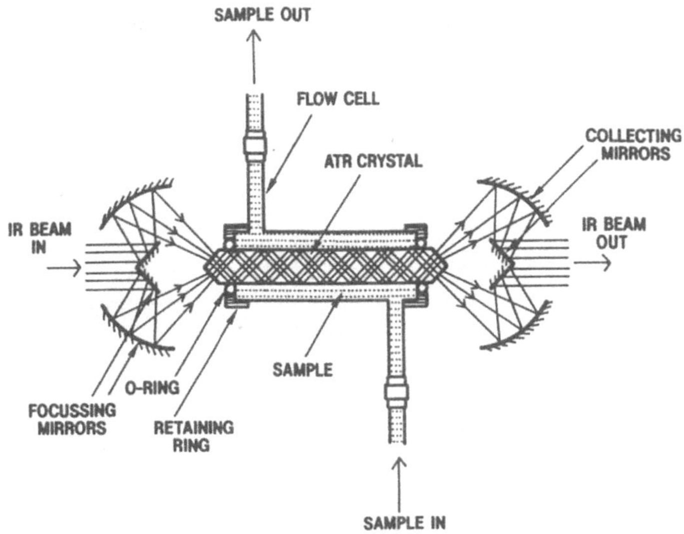

Figure 3.6 Schematic of a flow-through ATR cell employing a cylindrical crystal.

suffer from a restricted spectral range.

An example of a commercially available flow-through accessory is the Specaflow ATR system from Graseby Specac Ltd (Orpington, Kent). Based on a HATR accessory, it is designed to produce a laminar flow of sample over the ATR crystal, and can be used with mobile and viscous liquids at pressures up to 60 psi. Similar systems can be obtained based on the Spectratech (Warrington) Circle and Graseby Specac Squarecol cells, which employ cylindrical and square cross-section crystals, respectively. Other systems include ruggedized units able to operate at temperatures of up to 200°C and sample pressures of 150 psi.

Many samples which can be analysed using conventional FT-IR can also be successfully monitored using flow-through accessories. Proven applications include the determination of sugars and alcohol in fermentation broths (Alberti *et al.*, 1985) and of sugar mixtures in soft drinks (Kemsley *et al.*, 1992a). However, in spite of considerable adaptation for process situations, flow-through cells have in general been restricted to the laboratory or pilot plant, where they are likely to remain, until the spectrometers for which they are designed are equally capable of withstanding the harsh process environment.

An alternative sampling strategy is to use beam steering optics to connect the FT-IR spectrometer to specially constructed sampling cells, situated outside the spectrometer housing. There are advantages to this approach. The risk of damage to the interferometer associated with bringing fluids into the sample compartment is eliminated. The spectrometer can be sited well away from the process line, and protected against leaks and spills. The sampling cell can be mounted truly 'on-line' inside a pipe or reaction vessel, rather than plumbed into a spur taken off the main process stream. Light pipes may be used as the beam transfer optics, in conjunction with planar mirrors if corners need to be negotiated. A light pipe comprises a hollow metallic tube, the inside surfaces of which are gold coated to minimize losses. The AXIOT system of transfer optics from Axiom Analytical (Lagura Beach, California, USA) consists of a series of gold-coated light pipes and 90° rotary reflecting couplings. Each coupling is free to rotate about the axis of either of the connecting light pipes, allowing free articulation of the system. The diameter of the light pipes may match the size of collimated beam emerging from the interferometer. Alternatively, condensing optics may be used to reduce the beam cross-section, enabling the use of smaller diameter pipes. The beam transfer system is coupled to a sampling accessory. Generally, this is an ATR cell for use in the MIR; but NIR transmission cells with adequate path lengths are also feasible. The DPR series immersion probes from Axiom Analytical comprise an ATR element built into a probe-style housing suitable for coupling to the AXIOT light pipes, and are available in a range of sizes. In conjunction with a ZnSe element, the DPR probe has been used for monitoring the progress of chemical reactions (Doyle, 1989) and inspecting hazardous waste. However, for food applications, safety considerations may again impose restrictions on the choice of crystal material.

Fibre-optic FT-IR has attracted much interest as a method for remote sensing. Most instrument manufacturers now offer fibre-optic accessories of some kind, comprising an interface to the spectrometer, a length of fibre-optic cable, and a remote sampling cell. Some systems use fibre bundles, similar to dispersive NIR instrumentation. Others take advantage of the high throughput available from the FT-IR spectrometer, and use much cheaper single-core fibres. An optical fibre consists of a central core of glass surrounded by a cladding of a second glass of slightly lower refractive index. Light is transmitted along the core by internal reflection at the core-cladding interface, similar in principle to the operation of an ATR element. The whole fibre is sheathed in protective material. The core and cladding materials can be oxide, fluoride or chalcogenide glass. Oxide glass fibres (Figure 3.7) are the familiar telecommunications cables, and are widely available, robust and relatively cheap. Their transmission region is limited to the NIR. Core and cladding are generally based on low OH-content SiO_2 glasses. Fluoride glass fibres (typically ZrF_4) can transmit further into the IR, to around 2000 cm^{-1}, but they are more expensive and fragile. Chalcogenide glasses, defined as materials containing at least one of the elements S, Se and Te, have theoretical transmission windows covering the whole of the MIR. However, due to current manufacturing difficulties, chalcogenide fibres suffer from less than ideal attenuation

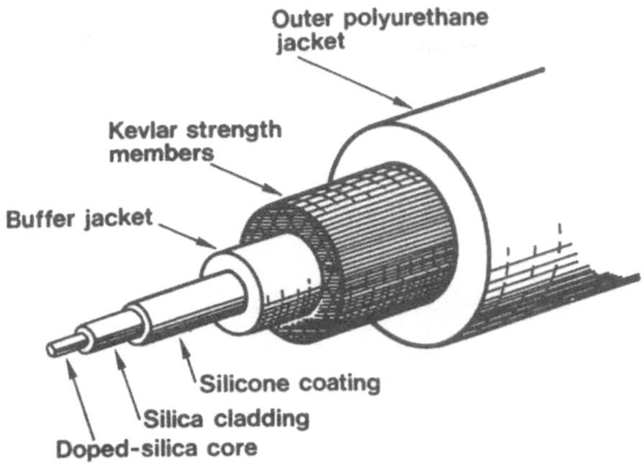

Figure 3.7 Section through an oxide glass optical fibre.

characteristics, and usable lengths are generally restricted to under ten metres. The fibres are also brittle, requiring substantial protection, and many are highly toxic.

A fibre-optic remote sampling cell may be a transmission type for use in the NIR, or an ATR type for use in the MIR. Samples are presently restricted to fluids and semi-solids. The cells are designed specifically for use with fibres, and are often in 'probe' form, such as the Fibreprobe range from Graseby Specac Ltd. Figure 3.8 shows a schematic of a typical fibre-optic FT-IR system. Such instrumentation is beginning to become established as a viable option for process monitoring, and applications such as the quantification of carbohydrates and ethanol in soft and alcoholic drinks have already been demonstrated (Kemsley *et al.*, 1992b). For food applications, toxicity considerations have until recently confined the use of fibre-optic FT-IR to the pilot plant. However, a novel fibre-optic sampling probe based on a sapphire ATR crystal is now available from Graseby Specac Ltd, and is suitable for use in the food processing environment. Proven applications of the probe include the quantitative analysis of sugar mixtures in solution.

Although fibre-optic FT-IR provides a convenient technique for remote sensing, a spectrometer coupled to a single fibre-optic accessory may offer no advantage over light-pipe remote sensor systems. This is particularly true in the MIR, where the optical fibres are expensive, and suffer from serious mechanical and optical limitations. However, if the interface can accommodate multiple fibres to link a number of sampling stations to a single spectrometer (Figure 3.8), the benefits may be significant. The relatively high cost of FT-IR spectrometers has hindered their establishment in the process field; a potential buyer needs to be convinced that their

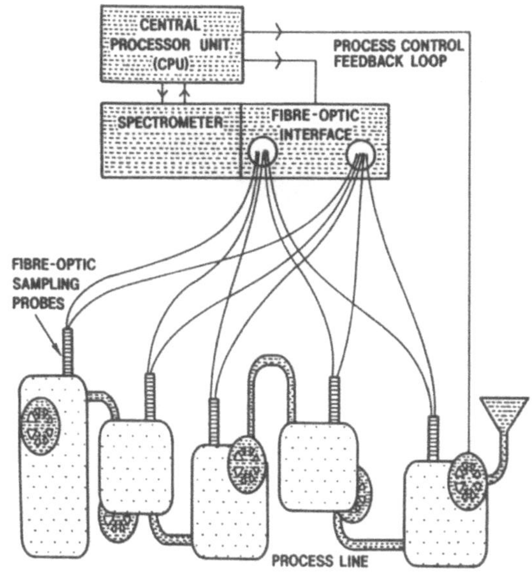

Figure 3.8 Schematic of a fibre-optic FT-IR system for process monitoring and control.

installation on a process will lead to a substantial increase in profitability, particularly if multiple instruments are required. Fibre-optic multiplexed systems can offer a cost effective solution, and are likely to accelerate the acceptance of FT-IR into the process industry.

3.3.3 Dispersive instruments

The length of time required to scan a complete spectrum, and the intrinsically low throughput, combine to make dispersive instruments impractical as process control devices in the MIR. In contrast, the throughput of NIR instruments is high. Brighter sources can be used, and detectors are more sensitive. Furthermore, the sampling techniques used in the NIR can much more readily be adapted for process use. Low-loss transmission cells, which incorporate silica optical fibres to achieve remote sampling, are widely used (Feldhauser *et al.*, 1989). In the past, the fibres have tended to be in the form of bundles, often shaped to match the slit geometry of the spectrometer, and these have been costly. More recently, improvements in fibre and spectrometer instrumentation have enabled the use of large diameter (300–500 μm), single-core fibres. For example, the Model 300 Mainframe system produced by Guided Wave, Inc (El Dorado Hills, California, USA) is an NIR dispersive spectrometer which used single-core optical fibres to interface up to eight separate sampling cells. Comprehensive software allows the fully automatic collection and processing of data from the multiple sampling points.

Although a wide range of samples in a variety of disciplines may be handled using this instrumentation, for food applications the potential of the technique is limited by the need for large calibration sets and complex statistical procedures invariably associated with NIR spectroscopy. For this reason, dispersive IR faces increasing competition from FT-IR, as the higher throughput and favourable optical geometry of the FT-IR spectrometer promises to enable fibre-optic spectroscopy, not only in the NIR, but throughout the spectroscopically more attractive MIR too.

References

Alberti, J.C., Phillips, J.A., Fink, D.J. and Wacasz, F.M. (1985) Off-line monitoring of fermentation samples by FTIR/ATR; a feasibility study for real-time process control. *Biotechnology and Bioengineering Symposium*, No. **15**, pp. 689–720.

Belton, P.S. and Wilson, R.H. (1990) Infrared sampling methods. In *Perspectives in Modern Chemical Spectroscopy* (ed. Andrews, D.L.). Springer-Verlag, Berlin, Heidelberg, pp. 67–86.

Brown, C.W., Lynch, P.F., Obremski, R.J. and Lavery, B.S. (1982) Matrix representations and criteria for selecting analytical wavelengths for multicomponent spectroscopic analysis. *Analyt. Chem.*, **54**, 1472–1479.

Doyle, W.M. (1989) Laboratory scale reaction monitoring using an FTIR deep immersion probe. *Proc. 7th Internat. Conf. on Fourier Transform Spectroscopy*, vol. **1145**, pp. 616–617.

Feldhauser, B., Meya, K. and Siesler, H.W. (1989) Near infrared spectroscopy with fiber optics and chemometric data treatment. *Proc. 7th Internat. Conf. on Fourier Transform Spectroscopy*, vol. **1145** pp. 158–160.

George, W.O. and McIntyre, P.S. (1987) *Infrared Spectroscopy* (ed. Mowthorpe, D.J.). John Wiley and Sons, Chichester.

Griffiths, P.R. (1975) *Chemical Infrared Fourier Transform Spectroscopy*. Wiley, New York.

Harrick, N.J. (1987) *Internal Reflectance Spectroscopy*. Harrick Scientific Corporation, New York.

Hyvärinen, T. and Niëmela, P. (1990) Rugged multiwavelength NIR and MIR analyzers for industrial process measurements. *Proc Conf. on In-Process Optical Measurements*, The Hague, March 1990, vol. **1266**, pp. 99–104.

Jones, R.W. and McClelland, J.F. (1990a) Quantitative analysis of solids in motion by transient infrared emission spectroscopy using hot-gas jet excitation. *Analyt. Chem.*, **62**, 2074–2079.

Jones, R.W. and McClelland, J.F. (1990b) Transient infrared transmission spectroscopy. *Analyt. Chem.*, **62**, 2247–2251.

Kemsley, E.K., Wilson, R.H. and Belton, P.S. (1992a) Potential of Fourier transform infrared spectroscopy and fiber optics in process control. *J. Agri. Food Chem.*, **40**, 435–438.

Kemsley, E.K., Li Zhuo, Hammouri, M.K. and Wilson, R.H. (1992b) Quantitative analysis of sugar solutions using infrared spectoscopy. *Food Chem.*, **44**, 299–304.

Martens, H. and Naes, T. (1989) *Multivariate Calibration*. John Wiley and Sons, Chichester.

Miller, R.G.J. and Stace, B.C. (1979) *Laboratory Methods in Infrared Spectroscopy*. Heydon and Sons, London.

Wilson, R.H. (1990) Fourier transform mid-infrared spectroscopy for food analysis. *Trends Analyt. Chem.*, **9**(4), 127–131.

4 Machine vision for the food industry

J.P. CHAN and B.G. BATCHELOR

4.1 Introduction

Rapid advances in the automation of production methods have increased inspection requirements for three main reasons. First, higher production speeds require higher inspection speeds. Secondly, the implicit inspection involved in manual production and assembly is no longer present and must be accommodated elsewhere and lastly, there is an ever increasing demand by the customer for higher quality. Machines that can 'see' have been developed for a variety of tasks that involve inspecting and manipulating industrial artefacts. This chapter discusses some of the problems faced by the food industry and possible solutions using machine vision systems. Machine vision is defined by the Machine Vision Association of the Society of Manufacturing Engineers as,

> 'The use of devices for optical non-contact sensing to automatically receive and interpret an image of a real scene in order to obtain information and/or control machines or processes.'

In other words, a machine vision system is an attempt to automate certain processes based upon its visual appearance, without any human interaction. The are two main application areas for machine vision systems; product and process inspection, and control of electromechanical devices (robot vision). Industrial inspection using automated image analysis is most commonly called automated visual inspection (AVI) (Bachelor *et al.*, 1985).

The motivation for the use of machine vision systems for inspection and process control can be summarised thus (Bowman and Batchelor 1986):

- Reliability
- Versatility
- Speed
- Integration of production processes
- Cost saving

Although no machine can match the qualitative and interpretative powers of human vision, simple repetitive tasks can be better performed by a machine vision system than a human inspector. Such a system is immune to boredom or fatigue, and can be far more consistent in the quality of inspection.

In a number of manufacturing processes, 100% inspection is imperative to prevent faulty and dangerous products from reaching the consumer. Under these circumstances, the opportunity of 100% inspection using AVI becomes an extremely attractive, almost essential proposition. Another advantage of AVI is that there is no contact with the item being inspected and no gauges to wear, so that reliability is enhanced and greater consistency assured.

In comparison to other forms of quality assurance (QA) techniques, visual inspection is very versatile. AVI has the potential to solve many inspection problems that are difficult to solve in any other way, e.g. examining surface defects such as scratches and cracks, or the measurement of non-linear features such as area and angles. Being a non-contact process, AVI has important implications in the inspection of delicate or hygienic products such as food and pharmaceuticals. Also, it can be used in harsh environments where manual inspection is not possible.

There are some products that simply cannot be inspected using existing techniques at the rate at which they are produced. In many simple tasks, a machine vision system can perform inspection at much higher speeds than a human operator can. In a number of AVI applications, sufficient speed has only been achieved by using dedicated hardware. However, advances in powerful, high-speed and low-cost microprocessors, along with the development of special image processing computer architectures, are greatly improving the speed of general purpose machine vision systems.

The fourth motivation for the use of machine vision systems is the integration of inspection with the production equipment. Provided the equipment is trouble free the reporting on defects is much closer to the event than is possible with manual systems. For example, if a baked product is detected to be overcooked by a vision system, feedback control can instantaneously reduce the temperature of the oven whilst a feed-forward loop can operate a reject mechanism. Swift and automated reaction to any defects means that machine, man and material utilisation are all improved.

The final and perhaps the most immediate motivation for a manufacturing company, is that of the cost savings which can be obtained by eliminating expensive manual inspection, especially in repetitive high-volume production. The cost benefits are becoming more attractive as the cost of computers, memory and solid state imaging devices falls. Image processing equipment that was in the mid-1970s, confined to research laboratories is now available at a fraction of the cost, many times the performance and just as importantly, in much more compact forms.

Although automated visual inspection has been applied in many areas, the food industry has so far been unable to take full advantage of this technology due to the variable nature of product appearance. The key to success in the 1990s for food manufacturers would seem to be a variation in ingredients and processing routes (Lightbody, 1989). It is sometimes desirable for food products such as cakes, pies, pizzas and chocolates to feature a significant, but controlled, amount of variability. Such variability is hard to quantify precisely and the faults associated with these products are also varied and difficult to predict. To compound the problem, there are pressures for product innovation, particularly on multi-product processing plant. To date, most vision systems have been designed for dedicated inspection tasks. There is a need for systems that are capable of adapting to new situations, perhaps automatically, but at least without requiring extensive reprogramming by experts.

4.1.1 Customers judge food by its appearance

The ever increasing awareness of the consumer, has led to demands for higher quality in all consumer products. This is especially true in the case of food, as stricter product liability laws will testify. Maintaining a high standard is imperative if a company is to survive in today's competitive environment.

In faults relating to products, there are two common problems faced by the food industry: 'badly made' products as a result of a manufacturing fault and undesirable foreign bodies introduced either during the making of the product, or in the basic ingredients prior to mixing.

A proportion of 'badly made' products is inherent in all high volume production lines, the causes of which may include incorrect settings, mispositioned feed, blockages, faulty machinery and human error. In comparison with badly made products, foods with foreign bodies occur at a much lower frequency, however they are potentially the most damaging. A badly made product can usually be spotted at the factory, whereas foreign bodies may go unspotted. Furthermore, a customer is more likely to react contemptuously when faced with a sliver of glass in his meat pie than finding a badly decorated pastry. Foreign bodies can be introduced into the foodstuff from a variety of sources:

- Unwanted objects in the base ingredients
- Parts of the machinery and/or operators
- Insects and/or rodents caught in the production process
- Parts of packaging used, such as glass in bottled products
- Industrial sabotage

The following is a selection of headlines that appeared in various newspapers within a space of a few months in 1989 (Chan *et al.*, 1990):

'Mouse in milk!'	22/6/89
'Glass in crisps'	20/7/89
'Butter recall' (metal pieces found in butter)	18/2/89

'More glass in baby food'	13/4/89
'Glass in meat tin'	30/7/89
'Remove these danger sweets' (glass fragments)	25/5/89

More recent claims to fame (in 1991) include 'Lizard in cornflakes', 'Bird's scull baked into loaf', and 'Razor blades in yoghurt'.

The range of faults that can occur in a food product is wide and varied, but they can be generally classed into two types: visible by naked eye, and non-visible. The use of machine vision for both these cases is possible by utilising sensing methods that respond to electromagnetic wavelengths beyond that of human vision.

Automated visual inspection offers an ideal solution to food inspection being noncontact, non-destructive and offering the ability to incorporate feed-forward and feedback loops to control the manufacturing process. AVI, however, has so far been limited to precision made products, and performing simple tasks like counting and segregation of more complicated objects. Complex tasks, like the inspection of surface decorations, are difficult due to the high tolerance of the product specification. The positioning of toppings on a pizza for example, can vary greatly and still be acceptable. Furthermore, the level of acceptability can be very subjective.

The advent of 'intelligent' vision systems, together with cheap fast hardware, has now enabled food companies seriously to consider the use of machine vision for inspection and monitoring tasks.

4.1.2 Human vision versus machine vision

Human vision is basically qualitative, it is very good for making judgements about the appearance of objects and the general quality associated with that appearance. It deals instantly with the global information in a scene, it can solve ambiguities due to perspective, lighting and amplitude. Current machine vision systems are considered to be better at quantitative tasks, such as helping to make a measurement and quantify a value. It is a mistake to dwell on the comparisons between machine and human vision, no machine vision system today or in the foreseeable future can approach the interpretative powers of human vision.

Machine vision has some properties that are better than human vision in that they can generally operate simple tasks at much higher speeds. Whereas people would have difficulties making consistent inspection judgements on objects running any faster than say 40 times a minute, there are machine vision systems today that can inspect up to 3000 objects per minute. In fact, at best a human operator can only expect to be 60-70% efficient under normal operating conditions. A machine vision system can, however be expected to perform objective and consistent judgements 24 hours a day, 7 days a week if required. It does not get tired, distracted or require to perform bodily functions, and it certainly never asks for a pay rise!

4.2 Machine vision systems

The level of interpretation and decision involved in human vision can be easily taken for granted, developing as it does on the basis of prior knowledge gained through trial and error. The design of a machine vision system requires a broad range of technical skills including system analysis, programming, hardware and software engineering, optics and a knowledge of digital image processing. Figure 4.1 is a block diagram illustrating the main components of a machine vision system.

4.2.1 Image formation

An image is a two-dimensional light intensity function $f(x,y)$, where x and y denote spatial coordinates and the value of f at any point (x,y) is proportional to the brightness of the image at that point. This brightness, in monochrome images, is also referred to as the grey level or grey scale value. A digital image is an image $f(x,y)$ which is discrete in both spatial coordinates and in brightness. The individual elements of a digital image

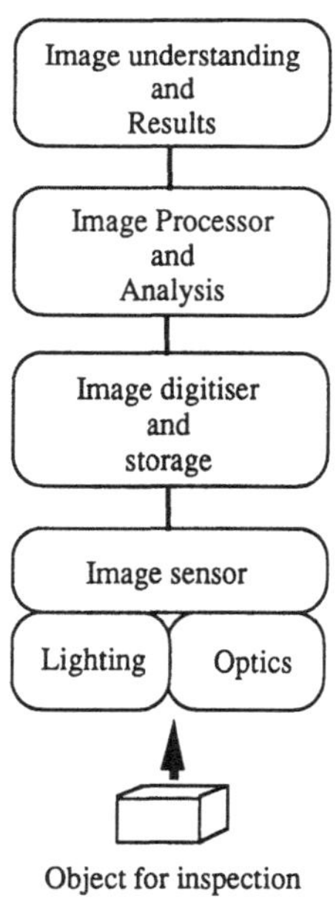

Figure 4.1 Basic components of a machine vision system.

are referred to as pixels. Image formation systems consist of optics, some form of illumination and a sensor (i.e. the camera). The system generates an analogue representation of the scene (e.g. a video signal).

It is critical that the image formation component of our vision system is properly designed. A common mistake is that not enough effort is spent on the image formation resulting in over-complex algorithms to overcome a problem that could have been solved by a simple change in lighting at the image formation stage.

4.2.1.1 Lighting techniques. In the majority of machine vision applications, the importance of lighting cannot be over-emphasised. The appearance of the an object is critically dependent on its attitude to the light source, as well as the colour, polarisation, spatial light distribution and physical size of the source. By adjusting the lighting to highlight or obscure features of the object being inspected, the image processing algorithms may be simplified and the whole system made more robust.

Factors to be taken into consideration when choosing a method of illumination include the following:

- Wavelength required
- Intensity required
- Area to be illuminated

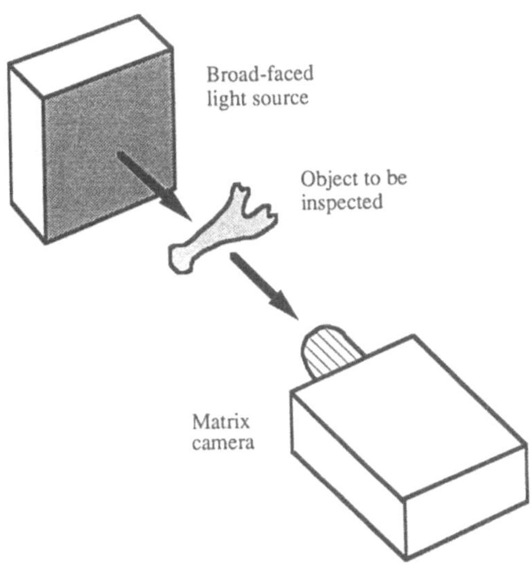

Figure 4.2 Back illumination using a matrix camera to view a stationary, matt, opaque object.

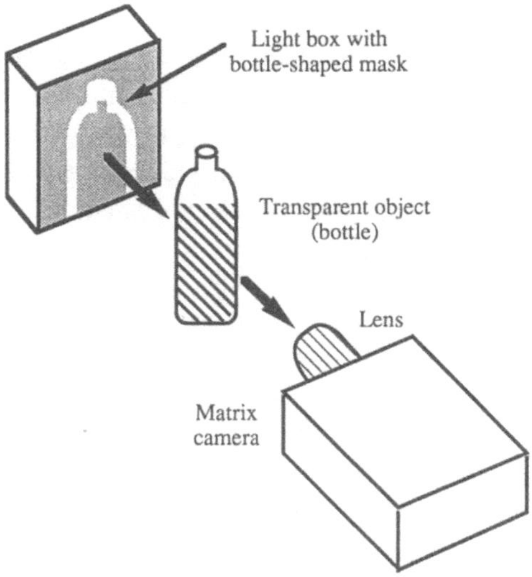

Figure 4.3 Dark-ground illumination using a matrix camera to view a transparent object.

- Reflectivity of object to be illuminated
- Transmission efficiency of image and acquisition equipment
- Space available

Figures 4.2 to 4.9 illustrate a number of common lighting techniques used in inspection (Batchelor *et al.*, 1985).

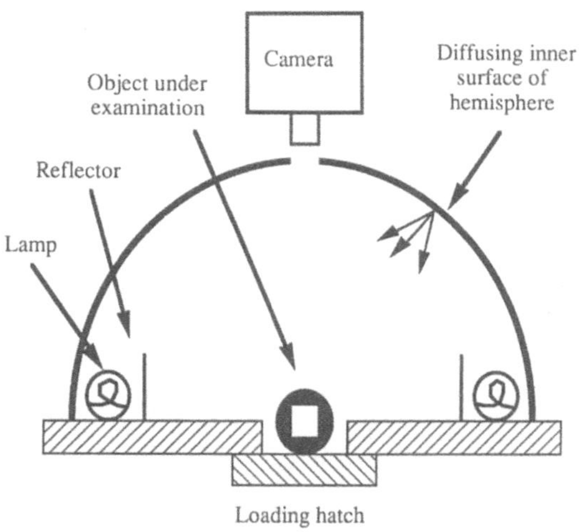

Figure 4.4 Omni-directional lighting.

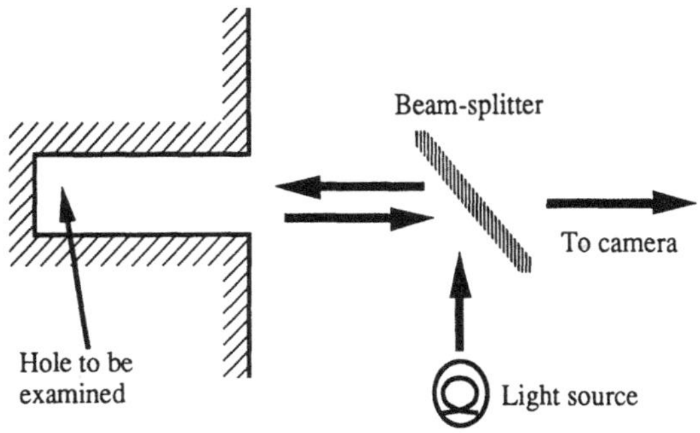

Figure 4.5 Using a beam splitter to view inside a hole without casting shadows.

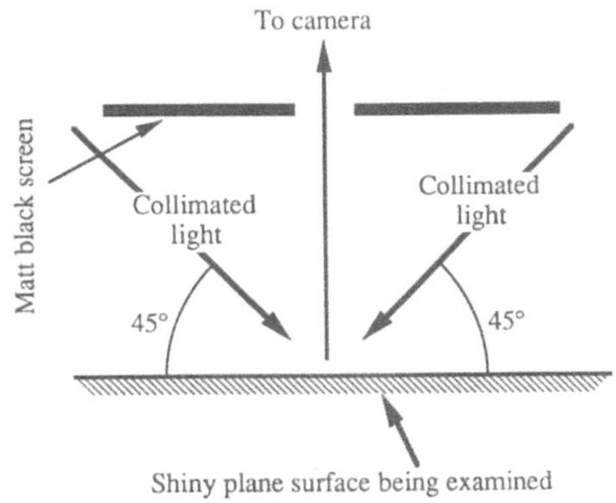

Figure 4.6 Avoiding the effects of glinting when examining a plane shiny surface.

4.2.1.2 Optics. The optical techniques used in machine vision systems range from the use of lenses or fibre-optic devices for illumination or image acquisition, to complex optical processing methods. Fibre optic and conventional optical devices are useful for 'extracting' images from inaccessible places, e.g. the use of endoscopes. Coherent fibre-optic bundles are useful for extracting images in situations where lack of space or harsh environment conditions restrict the use of a camera.

The careful design and use of optics can greatly simplify a machine vision task, resulting in a cheaper, faster and more robust system. For example, to separate red tomatoes from green tomatoes, it is not necessary to involve expensive colour cameras or complex image processing algorithms. A simple green filter would make red seem black and green seem white. It then becomes a trivial task of distinguishing black from white!

Optical processing methods show great promise for AVI because of the power and speed. Optical techniques available include holography, correlation and diffraction. A particularly useful optical technique is the ability to perform two-dimensional (2-D) Fourier transform in negligible time by

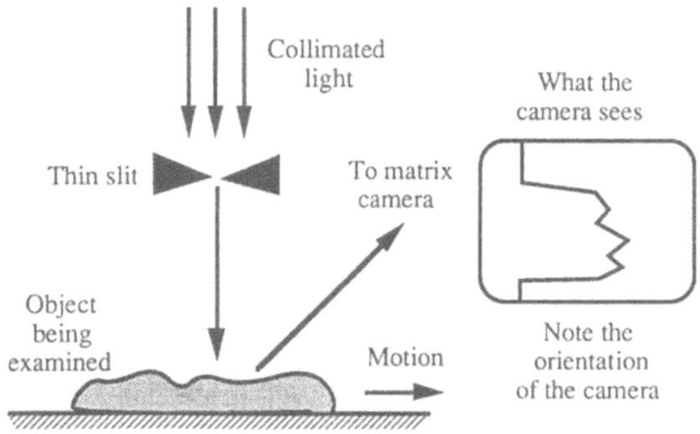

Figure 4.7 Structured lighting for 3-D shape measurement.

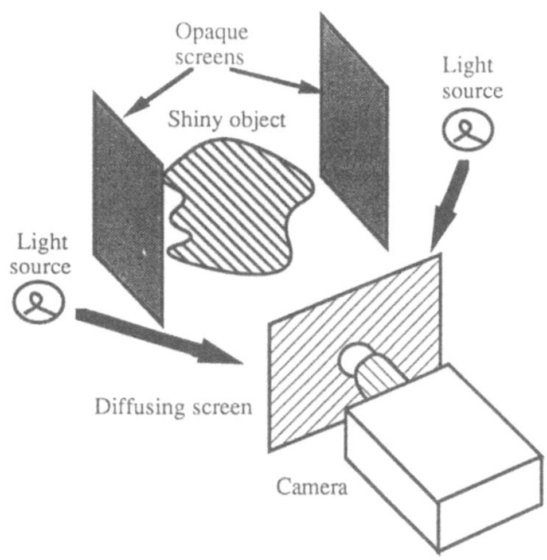

Figure 4.8 Using a diffusing screen to illuminate a shiny object.

forming diffraction patterns (Figure 4.10). A laser and beam expander produce a coherent wave front which illuminates the object; in this case a transparency. The 2-D transform is formed at the diffraction plane by using L1 to focus the light diffracted by the object. The inverse transform is performed in a similar manner using L2. A filter mask can be placed at the diffraction plane to block out certain sections of the diffraction pattern and the corresponding filtered image appears instantaneously at the image plane. This method is ideally suited to inspecting surfaces with regular features or geometry.

Another example of optical technique is the use of Moire fringe patterns for producing accurate depth data (Reid *et al.*, 1984).

4.2.1.3 Image sensors. The role of the image sensor is to transform

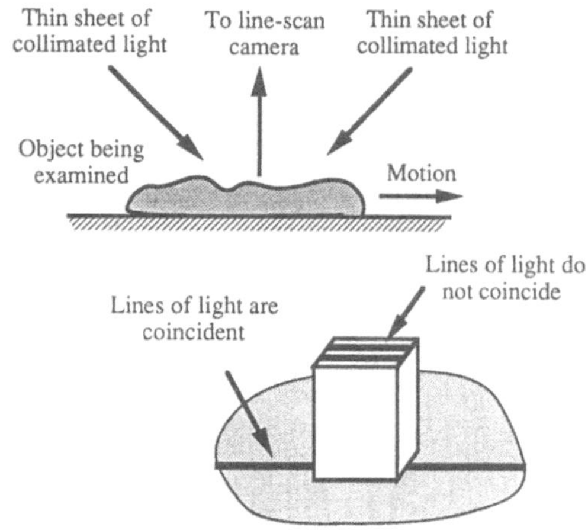

Figure 4.9 'Consight'. A method for recognising the presence of an object. It does *not* measure 3-D shape.

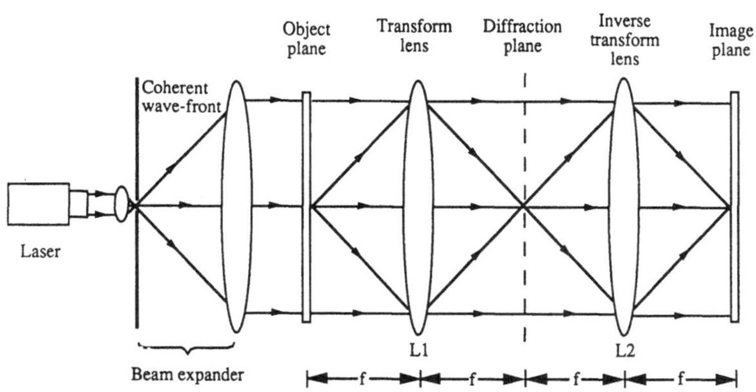

Figure 4.10 Optical 2-D Fourier transform.

certain information related to the physical make-up of a scene into light intensities that can be viewed by the human eye and/or digitised by the vision system. The most common form of image sensor used in visual inspection systems is the video camera, which is sensitive to visible light reflecting off surfaces in its field of view. Its end result is to emulate the perception of the human eye. Here we will briefly review the types of video cameras available.

CRT cameras. Cathode ray tube cameras (CRT) were widely used before the arrival of charge couple device (CCD) cameras and they are still used in specialised applications such as situations where there are low lighting levels, extremely fast moving objects and where the camera is exposed to high radiation levels. A wide range of tubes is available, each with their own characteristics. The most common tube for general purpose use is the low cost Vidicon. Other tubes in use include the Newvicon (greater sensitivity), Plumbicon (low lag, low noise and high sensitivity), Chalnicon (high sensitivity, wide spectral response, resistance to image 'burn in'), and Pyroelectric Vidicon (sensitive to infrared wavelength).

Solid state array cameras. These are probably the most common method of image sensing in use today. These cameras contain arrays of photosensitive elements fabricated on a single chip. The technology used is usually termed charge couple device (CCD). Solid state cameras have a number of advantages over CRT cameras:

1. The sensors are smaller and lighter than a CRT tube.
2. They are robust, resistance to minor knocks, vibrations.
3. The picture geometry is fixed at manufacture and remains stable, unlike a CRT, which is likely to drift due to age.
4. A CRT is affected by external magnetic fields, whereas CCDs are not.
5. The power consumption is much lower than a CRT.

Line-scan cameras. Line-scan cameras, or linear array sensors, utilise the same technology as solid state array cameras but are available with much higher single line resolution (up to 8192 × 1 elements). This means they can be used to good effect in applications requiring a high resolution, e.g. for dimensional checking or for accurate positioning. Line-scan cameras can be useful in the inspection of moving objects, using the motion to scan along the second dimension.

Colour cameras. Using the technologies described above (CRT and CCD), colour information may be obtained by sensing the 'redness', 'greenness' and 'blueness' of the scene separately, and then combined to form a colour image. There are many different ways of representing colour video information, some of which are as follows: composite video, RGB (red, green, blue), HSI (Hue, intensity and saturation) and YIQ (luminance, chrominance amplitude and chrominance angle). Colour cameras can be very useful in the segmentation of objects by their colour, a simple example of isolating cherries on small round cakes is given in Figure 4.11. A RGB image consists of three components, by selecting just the green channel for example, red objects appear dark and green objects appear light (Figure 4.11(a)), whereas the red channel does the opposite (Figure 4.11(b)). Subtracting one from the other (colour difference), enables the cherries on these Bakewell tarts to be separated from the rest of the cake. The use of colour optical filters can give the same results, but would

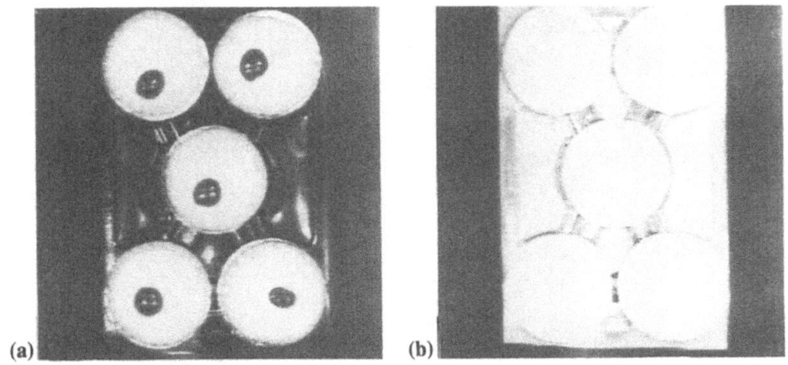

Figure 4.11 Using RGB information to enhance certain features of a product: (a) the green channel, i.e. the 'greenness' of the scene, (b) the red channel showing the redness of the scene.

require mechanical devices to apply and remove these filters, it would also mean a slowing down of inspection speed.

4.2.2 Non-visual sensing methods

Video cameras respond to visible light, emulating the perception of the human eye. There are many types of sensors which do not follow this emulation, like laser scanners, X-ray image intensifiers, thermal imaging cameras, and ultrasonic scanners (Perry, 1990). These sensors respond either to electromagnetic wavelengths outside that of the human visible range or to sound waves. This information is then transformed into an image which can be seen by the naked eye. This image can then be digitised and processed as a normal video image. Some of these sensing methods are discussed in this section. A review of the applications of non-visual sensing methods in the food industry is given later in this chapter.

X-rays offer promising applications for machine vision, giving information about the internal characteristics of an object being viewed. It has already been used quite extensively in the field of non-destructive testing, for example, inspecting steel welds for cavities. X-ray imaging of food products is not only useful for foreign body detection (Figure 4.12), but can also be used to check parameters such as texture and ingredients.

Gamma rays are similar to X-rays, but are generally much less expensive to produce, and can have much higher penetration power than X-rays. The drawbacks are that the images are of low contrast compared with X-ray images, and more importantly, there are additional and constant safety requirements due to the fact that a gamma ray source cannot be switched off.

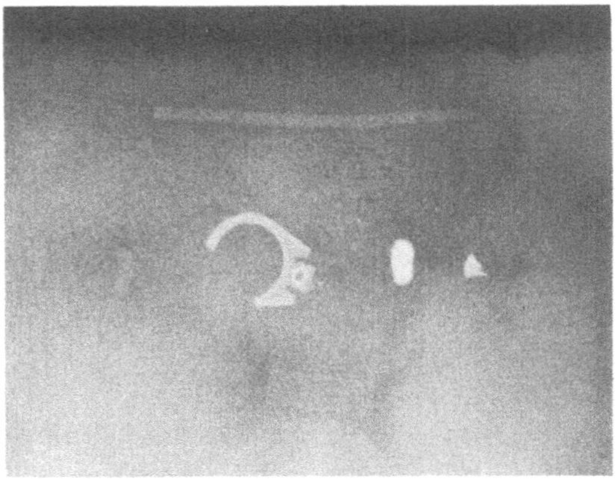

Figure 4.12 X-ray image of offal (a constituent of granary flour), with foreign bodies, including rust flake, stone, pull-can ring and elastic band. Image taken at 24 kV, 14 mA.

All substances absorb electromagnetic vibrations, usually over a characteristic range of wavelengths, and many emit or re-emit such radiation. This emission is known as luminescence. When luminescence lasts only during the period of excitation (using ultra-violet light for example), it is known as fluorescence. The use of ultra-violet fluorescence could be very useful in showing characteristics of a food product that would go undetected by the visible part of the electromagnetic spectrum, since the intensity and colour of the light emitted are characteristic of the material under investigation.

Ultrasound testing utilises the same principles as sonar, as used in sub-maritime exploration. This method may be used to provide range information for three-dimensional objects. Like X-ray, ultrasonic inspection has been used mainly in the medical field, and has proved to be a relatively safe method (unborn babies are examined using ultrasound). The main inhibiting factor to its use in industrial inspection is that it requires contact with the medium it is examining. Furthermore, a typical ultrasound image contains a lot of noise and even for a skilled operator, the interpretation of the image tends to be subjective.

Infrared imaging could be used for the detection of temperature gradients, for example hot or cold spots during cooking or freezing operations. Other useful imaging methods include nuclear magnetic resonance (NMR) and microwaves.

4.2.3 Image acquisition

Digital image acquisition equipment is concerned with the generation of a two-dimensional array of integer values representing the brightness function of the actual scene at discrete spatial intervals. This analogue video signal is fed to a digitiser or frame-grabber where it is sampled at discrete intervals to give a matrix representing the image. The size of this matrix can vary between 128 × 128 and 1024 × 1024 elements or pixels, depending on the resolution required. The brightness function is also quantised to give the grey scales, typically 256 levels.

The ability to process data efficiently and purposefully depends on the nature of the encoding used to represent the original image, hence the importance of image storage. Images may be encoded and stored in different ways depending on the type of image required. The following are common types of images used in machine vision:

- Grey scales images (monochrome);
- Binary images (i.e. black or white);
- Colour images, which can be represented by three separate monochrome component images;
- Image sequences (representation of moving pictures).

4.2.4 Image processing

No matter how well the image formation/acquisition performs, the image in the framestore is invariably not good enough to be analysed directly. When this occurs, the image has to be processed. The term image processing is taken here to describe techniques that transform one image into another, so that the properties of the image to be analysed are enhanced. This can be done by the removal of certain properties in the image that could confuse the image analysis. For example, if the image is 'noisy', it is often useful to transform the image through a low-pass filter to eliminate the noise. Similarly, it may be desirable to enhance certain features in an image at the expense of other features. A good example of this is edge detection, this is a process in which the original grey scale image is transformed into an image where the edges or outlines of objects are highlighted (Figure 4.13). Other common image processing techniques include:

- Thresholding (Local, global and dynamic)
- Image segmentation
- Contrast stretching
- Histogram equalisation
- Filtering (high-pass, median)
- Skeletonisation
- Convolution
- Hough transform
- Fourier transform

Image processing operations can be generalised according to the following schedule (Batchelor, 1991):

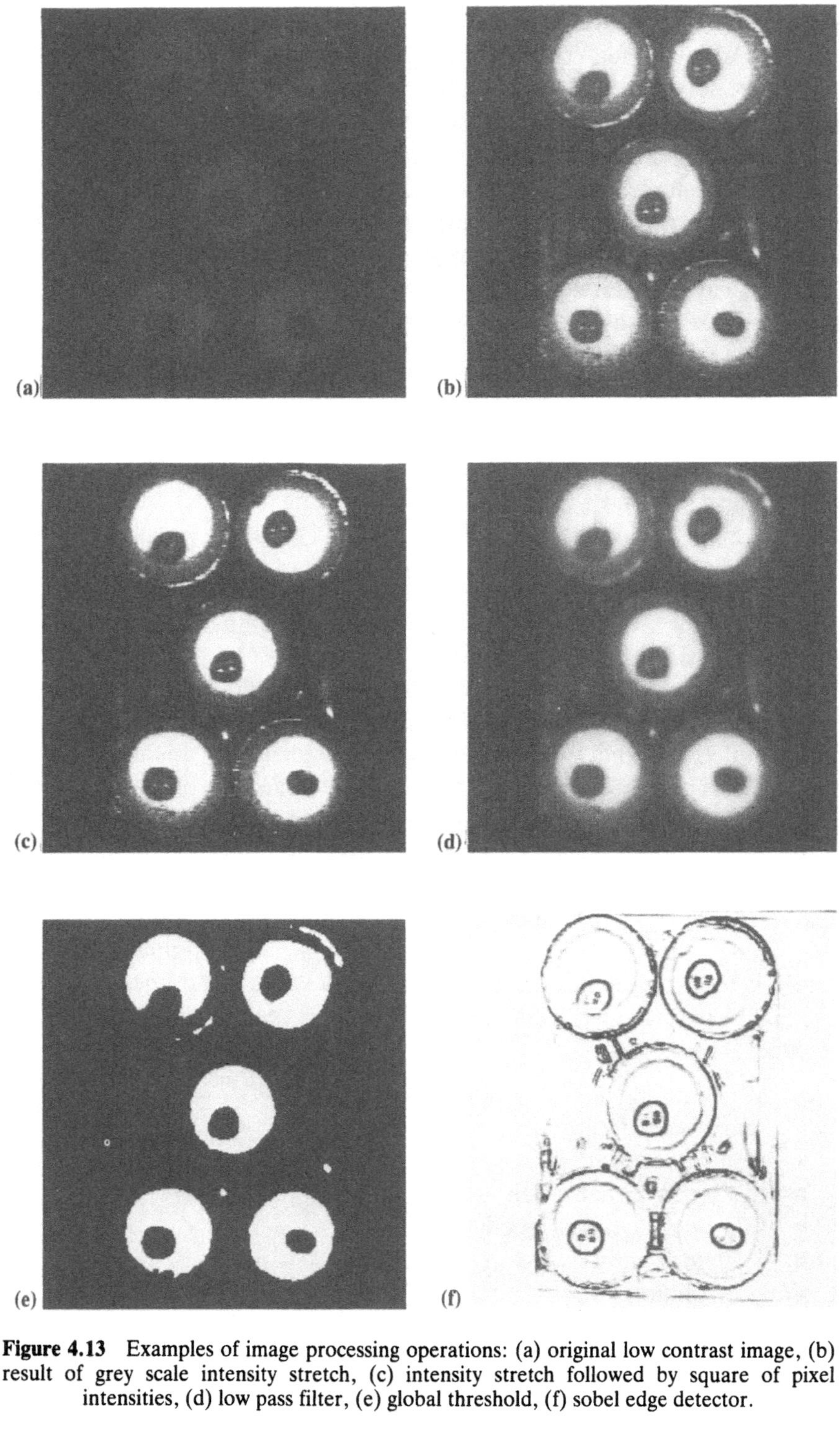

Figure 4.13 Examples of image processing operations: (a) original low contrast image, (b) result of grey scale intensity stretch, (c) intensity stretch followed by square of pixel intensities, (d) low pass filter, (e) global threshold, (f) sobel edge detector.

1. *Monadic point by point operators*, in which there is only one image and the resultant point, or pixel, depends on the same point in the input image.
2. *Dyadic point by point operators*, in which the resultant pixel intensity in the output image depends upon the intensities of corresponding pixels in two input images.
3. *Local operators*, in which the intensity at one point in the output image depends upon the intensities of a clustered group of pixels in a single input image.
4. *Global operators*, in which the intensity of at a given point in the output image depends on the intensities at a large number of points in the input image.
5. *Data dependent operators*, in which the function which is applied to an image is determined in part by the properties of the output image.
6. *Image transformation*, which results in some translation, rotation or warping of the input image.

4.2.5 Image analysis

Image analysis is a term used for techniques that transform an enhanced image into some form of description of that image. The output from an image analysis system are therefore not images, but descriptors, numbers or decisions. In order to be able to distinguish images it must be possible to generate a way to describe these images in terms of some pre-defined parameters. These parameters are also referred to as image features.

The three most common techniques used in image analysis are (Ballard and Brown, 1982; Whelan, 1991):

- Template matching
- Pattern recognition using feature extraction
- Descriptive syntactic processes

Template matching. In template matching, a prototype of the desired object is stored in memory. This is an ideal representation of the object to be found in the image. The template matching technique involves the translation of the template to every possible position in the image and finding a measure of the match between the template and the image at that position. If the match is within a specified range then the object is assumed to be present. In practice it is very difficult to match images repeatedly to the accuracy necessary for this approach to be considered useful.

Pattern recognition using feature extraction. There are essentially three components to pattern recognition using feature extraction:

- Segmentation
- Feature extraction
- Classification.

Segmentation is the process of partitioning the image into 'meaningful' components. It is a grouping process which identifies regions in the image as being similar with respect to some criterion. Segmentation algorithms are usually based on either similarity or difference of grey values, i.e. region growing or boundary detection. With the similarity approach the image is partitioned using thresholding, region growing and region splitting and merging. The boundary detection approach is based on abrupt changes in grey level.

It is usually desirable to represent the segmented regions in a form suitable for further processing. For example a boundary region may be represented by using a chain code, polygonal approximation or a signature. A homogeneous region may be represented by its skeleton. This data is now passed to the feature extractor.

The segmentation process results in a number of objects being found which are to be classified or identified. The next phase of the pattern recognition process is the extraction of features that are characteristic of the object and which will be used in the classification process. A feature may be described as a measurement that distinguishes or discriminates between a number of classes, e.g. the area of an object, length of object boundary. Feature extraction can be the most crucial and most difficult task in the image analysis system. This stage can often be greatly simplified if careful thought is given to the lighting and mechanical design. Some examples of common features are the length of a contour, diameter of a boundary, shape number of a chain coded boundary, area of a region or perimeter of a region.

Classification is concerned with the relation of the feature measurements to the actual object. An object is classified as belonging to some group or class on the basis of the features extracted in the last section. The features are combined to form a feature vector. A feature vector is defined as a set of measurements which is supposed to condense the description of relevant properties of the image into a small feature space. Objects of the same type will have similar feature vectors and thus will form clusters within the feature space. The most commonly used classification rules are the nearest neighbour classifier and the maximum likelihood classifier (Batchelor, 1979).

The choice of image features can greatly affect the performance of the classification process. For example, a feature such as a grey scale may be used in a tomato grading application but this feature may not be useful in an application, such as can end inspection, that requires size measurement. Generally, the features used in the classification process should have four characteristics (Castleman, 1978):

- Discrimination. Features should have significantly different values for different objects.
- Reliability. Features should have similar values for all objects of the same class. Tea that is found in a tea bag differs visually from ordinary leaf tea, but both belong to the same class, that is they are both tea. Therefore a feature that is chosen to classify tea from grass must be robust enough to account for the existence of the two grades of tea within its class.
- Independence. If possible, the various features should remain uncorrelated with each other.
- Small feature set. Since our classification task becomes more complex if a large number of features are used, it is wise to reduce the feature set as soon as possible in the development cycle.

In practice, the choice of features is largely intuitive, and the features selected do not always follow the suggestions outlined above.

If a feature falls within a certain range, then that image can be placed within a given group, this process is called the classification process. A simple example of this would be the grading of eggs by size; if after thresholding, the white pixel count (area) is used as the feature, then the eggs can be classified by saying that any egg that has a given area range belongs to a class known as 'big' eggs, whereas an egg belonging to a different range will belong to a class known as 'small' eggs, see Figure 4.14. On the introduction of the classification cut-off point into this example, the egg samples can be divided or classified into two groups, all the odd samples belong to class I, big eggs, while the even samples belong to the small eggs class, class II.

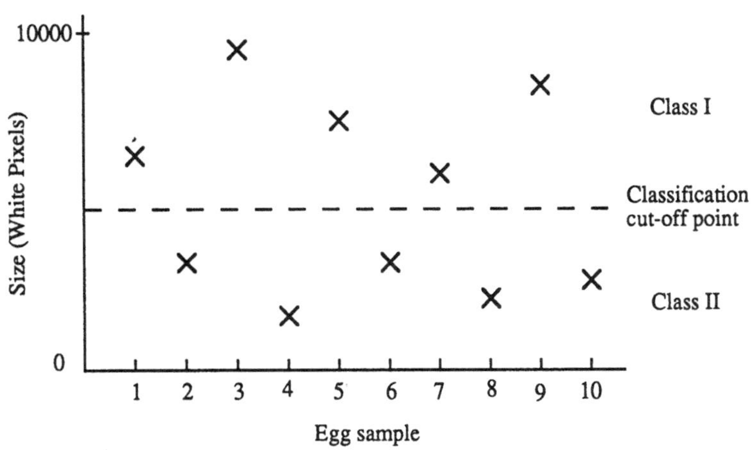

Figure 4.14 Egg classification example.

Descriptive syntactic processes. Unlike the pattern recognition techniques, which assume that the image contains one or more objects which are classified as belonging to some group based on one or more characteristic features of the object, the syntactic approach models an object by a set of features or structural elements, and by the spatial relationships between these features. This process requires a complete set of unambiguous rules from which we can gain an understanding of the scene to be analysed.

Consider the simple image shown in Figure 4.15. If we represent the horizontal lines in the image by the structural element 'a', and the vertical lines by the structural element 'b', then the syntactic representation of this image is the sequence ababa.

4.2.6 Image interpretation and understanding

Image interpretation is concerned with making some decision based on the information extracted from the image analysis section. This is, in effect, a further data reduction technique as the pictorial features are reduced to a smaller set of descriptors or even a single descriptor, covering those features. For example, in a scene of natural vegetation, we could describe the scene in terms of branches, leaves, fruit and flowers. Image understanding requires greater machine 'intelligence' than the previous stages and may incorporate high level concepts such as artificial intelligence (AI), declarative programming, neural networks, and expert systems (Chen, 1989). Image interpretation and understanding is becoming more and more important, as machine vision systems are applied to increasingly complex and variable scenes, such as those encountered in the food industry.

4.2.7 Current machine vision systems

There is an abundance of image processing hardware available 'off-the-shelf'. At the most basic level digitisers are available to plug into most desktop computers. These devices convert an analogue video signal into a digital one which the computer can store and manipulate. Digitisers which contain their own processing capabilities by way of an on-board central processing unit (CPU) are generally known as intelligent framestores. They may contain image processing functions already programmed in and

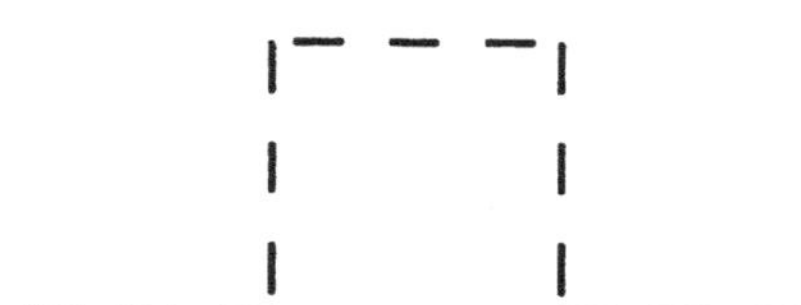

Figure 4.15 Syntactic representation example.

can be accessed in an interactive manner. Specialised image processing systems are also available, using sophisticated computer architecture such as pipeline and parallel processing, to speed up image processing operations.

The image processing (IP) hardware itself does not constitute a machine vision system. It merely provides a platform from which machine vision applications may be performed. Most suppliers of image processing hardware will also supply a library of IP routines, which can be included in a user's program for convenience. These routines are usually low level primitives for accessing the hardware and for manipulating and processing the image. Writing a IP application program is by no means easy, it takes a great deal of experience of what methods are suitable in order to enhance and finally analyse an image. Even with this knowledge, it is invariably a case of a 'try it and see' approach that finally decides on the algorithm used. This is one of the reasons why the development cost of a machine vision project is usually large compared with the components costs alone. Interactive image processing is a technique which helps to reduce the amount of effort spent developing an application (Batchelor, 1979). Interactive IP consists of a shell from which all image processing functions can be called instantaneously, and stringed together in some sequence if necessary. The example given here is Prolog + (Figure 4.16).

Prolog+ is intended as a prototyping tool for developing intelligent image processing and machine vision software (Batchelor *et al*, 1988; Chan and Batchelor, 1990). It is a multiprocessor system, designed to allow the integration of AI programming, image processing, expert systems and robotic/device control in a user friendly environment. The system architecture

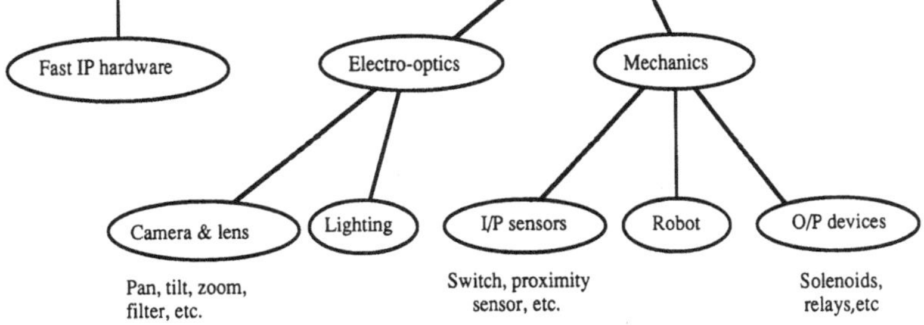

Figure 4.16 Prolog + control structure.

also permits processing of information from a number of sensing devices all of which are controlled via Prolog in the host computer. Prolog+ is an extension of Edinburgh Prolog that enables image processing and robot control commands to be included within the structure of a Prolog program. Functions may be implemented on pull-down menus and dialogue boxes. An interactive mode is also provided for similar to that of a command line interpreter.

As well as interactive image processing, other design tools exist that can help in the design of a machine vision application, they include lighting advisors (Batchelor, 1989), camera advisors and even image processing advisors (Chan *et al.*, 1991).

The development of a machine vision system is a difficult and costly task. Any tool that can help in reducing that effort is to be welcomed. The trend is for IP suppliers to provide hardware systems that can be programmed for specific applications with minimal effort. It is envisaged that future systems may include high level concepts such as natural language interface, speech recognition and synthesis, graphical user environment and graphical programming techniques. This would enable a system to be tailored to perform specific tasks without lengthy programming and design efforts.

4.3 Considerations for use of machine vision systems

4.3.1 Practical implications in using machine vision systems

In an automation project the cost of the vision system may be relatively small compared with the total cost and overall technology content of the project. The machine vision investment however, should not degrade the manufacturing process as a whole. It is therefore important that the implications for the other complex equipment are fully considered, when machine vision is augmented in the system. In order to ensure that no such degradation occurs, the following questions should be asked and answered as far as possible:

- Will the vision system effect the production speed?
- Will the production process have to be modified in order to accommodate vision?
- Will the production line have to be modified physically in order to provide a suitable station within the manufacturing line at which the vision equipment can be mounted?
- As the production demands change can the vision system be easily reconfigured?
- Is the vision system the least reliable piece of equipment in the line? If it is, will the introduction of the vision system reduce the line throughput?

- Can the service schedule of the vision system be determined accurately and can this be accommodated by the overall line service schedule?

A company can spend a lot of money and time introducing a vision system into a production line without fully considering the above questions, with the result that as soon as something changes in the production process or the vision system needs an unscheduled service, the line operators will simply shut down the vision system.

Machine vision is not a perfect inspection/measurement tool and we must be aware of the practicalities of a given application as well as the ultimate aim of the inspection. In any machine vision system there are essentially two main sources of error; system errors and statistical errors.

System errors are those that are inherent to the machine vision system itself. These include:

- Variations in the system lighting due to the non-ideal nature of the light sources.
- Distortions and variations in the system optics.
- The degradation of image quality due to the errors inherent in the image sensor. The errors introduced by the system electronics.
- Mechanical errors such as product mispositioning and variations in the background of the product.

Statistical errors are caused by the interaction of the machine vision system and the industrial process to which it is applied.

When a vision system is set up to accept certain objects and to reject others, there is a small 'grey region' where transition between good and bad occurs. This is the region where good objects get rejected and bad objects are accepted which leads to errors in the system's performance. This source of error is tied in closely with the machine vision system's interaction with the actual production process. The tighter the control the manufacturer has over the production process the smaller these errors become. In situations where it is not always clear where the division lies between a good and a bad product, as in the case of many food products, this uncertainty will manifest itself in terms of statistical errors.

The effectiveness of any system is bound by how it is used, this is no different for a machine vision system. Therefore the incorrect use or calibration of a vision system will result in the biggest source of error—it will not work!

It is clear that many factors have to be taken into consideration before implementing a vision system. There is no 'perfect' system, only what is reasonably effective at an affordable price. Invariably, the system will be a compromise of conflicting factors such as cost, speed, reliability, safety, flexibility and 'intelligence'.

4.3.2 Economic considerations in using machine vision

The task of justifying the inclusion of a machine vision system in an industrial environment is not always an easy one. It can take between 12 and 18 months of internal investigation before a company will consider investing in a technology such as machine vision, which many perceive to be new and unproven (Batchelor *et al.*, 1985).

When it comes to assessing the economic viability of a proposed machine vision system a number of factors should be considered. Besides the more obvious benefits associated with replacing a human operator or inspector, such as salary and training costs, there are some more intangible benefits.

When machine vision is used in an inspection and/or measurement role, a company can gain the advantage of increasing the quality of the goods they are shipping while at the same time reducing the costs of the goods sold. Since the cost of quality ranges from 15 to 35% of the total cost of the goods sold there is a significant opportunity for savings:

- Savings resulting from increased reliability of output.
- Maintaining your competitive position.
- Feedback from the vision system may be used to control the manufacturing process, thus improving product quality and reducing waste.
- Incipient faults can be recognised and corrected quickly, saving cost of wastage.
- Benefits from adding increased flexibility to the automation process.
- Automatic record keeping and reporting of production trends.

4.3.3 Environmental protection and safety

When a decision is made to implement machine vision into a manufacturing process it is important to consider the protection and safety of equipment, personnel and product.

Protection of equipment. A hostile factory environment is hardly an ideal place to house delicate electronic and optical equipment. It is therefore essential that the vision equipment is protected against all possible hostilities. A typical food processing plant could have extreme temperatures that vary with time as well as spatially: from refrigerated areas or 'quick-freeze' points to high temperature ovens. Other hazards include moisture, dust and electrical 'spikes'. The use of protective housing is a preferred option, but often means increased cost and at worst, compromising the performance of the system. Figure 4.17 shows how a camera could be housed in such a way as to protect it from a hostile environment. The use of fibre-optics for image formation may

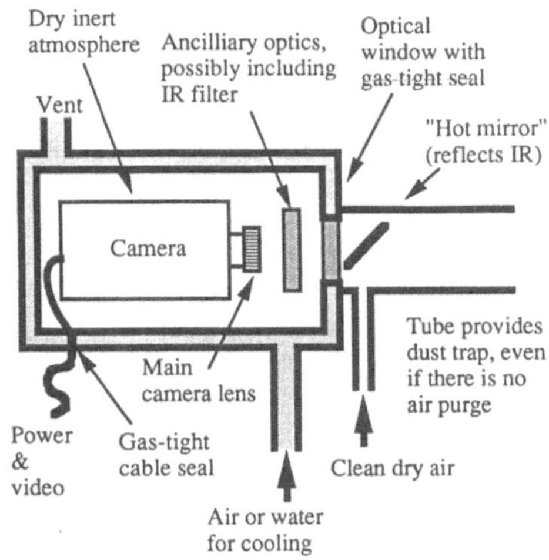

Figure 4.17 Protecting a camera from dust, fumes, heat and moisture.

also be useful. A variety of fibre-optics and optical accessories which enable good images to be obtained from inaccessible or hostile environments are available. This problem is inextricably linked with the problems of lighting, so that fibre-optics could also be useful in providing illumination in these situations. 'Ruggedised' versions of most popular desktop computers are readily available off-the-shelf and companies exist that specialise in 'ruggedising' equipment for factory use. Lastly, such expensive and finely calibrated equipment should be installed and housed in such a way as to minimise the possibility of tampering, accidental or malicious.

Protection of personnel. The protection of personnel from machinery and working environments is usually covered in the Health and Safety Act. If a vision system is to utilise hazardous methods, the design and implementation will have to be carefully considered. The use of stroboscopes, for example, could cause epileptic fits, migraine and serious accidents due to moving machinery appearing stationary. The use of X-rays or gamma rays for inspection has to be confined and personnel shielded adequately. Excessive dosage of X-rays or gamma rays can lead to anaemia, bone marrow diseases, genetic disorder and even leukaemia. It is essential that periodic checks are performed to ensure that leakages do not occur and that radiation does not exceed normal limits in the vicinity of the system. The type and degree of protective shielding depend on the power of the X-ray tube (or gamma source), with the area directly opposite the tube (facing the direction of X-rays) heavily shielded with lead plates. Ultraviolet illumination and high power lasers are also hazards that personnel has to be protected from. Exposure to them can lead to skin cancer and blindness, respectively.

Protection of product. As discussed earlier, it is possible to contaminate a food product from the production equipment, if the equipment is not maintained properly. Checks should be performed periodically to ensure that equipment is mounted and housed correctly, with no loose bolts and nuts and so on. For hygiene reasons, the equipment must be easy and safe to clean. This should be an integral part of the system design. The lighting equipment could pose serious hazards as most light sources contain glass and are of a fragile nature. It is easy to imagine a scenario where a light bulb shatters, contaminating the food products with glass fragments. There are two possible solutions to this problem: house the light units within transparent and shatterproof material such as polycarbonate or make extensive use of fibre-optics, so that fragile light bulbs may be located well clear of the products being inspected. Electrical and mechanical equipment are subject to safety requirements as would normally be expected in a factory environment.

Servicing. During the operation of the system, it is important to follow a few basic rules which could prolong equipment life as well as maintaining safety and performance:

- Keep equipment free from dust and moisture (unless the system is specifically designed to do so).
- There should be plenty of ventilation for all the hardware.
- Protect electronics from any electromagnetic interference. For example the image processing sub-system should be kept well away from a X-ray source.
- Protect the system from mains 'spikes'.
- Ensure the equipment is securely and correctly housed.

In order that the safety and performance of a vision system do not deteriorate, it is essential that regular servicing is carried out. To minimise the effort of this task, the system should be designed in such a way so that:

- Bulbs can be changed easily.
- System can be calibrated easily.
- System can be maintained with minimal effort.
- If possible, a self diagnosis of system faults is in operation.

Some components will have to be replaced after a period of use, for example an X-ray tube needs to be replaced every about three years, depending on the frequency and mode of use. When a component reaches the end of its useful life, it must be replaced promptly so as not to impair the performance of the system as a whole.

4.4 Machine vision applications for the food industry

4.4.1 Introduction

The use of visual sensing is potentially an ideal method for inspecting many foodstuffs since contact test methods can easily damage or contaminate them. Machine vision can also be fully integrated into a modern manufacturing environment where automation is increasingly the key player in reducing costs. As can be seen, the range of things that might be visually tested is immense and therefore so is the range of sensing and inspection techniques needed to solve these problems. Furthermore, the use of machine vision is not limited just to inspection of products it can also be used to monitor processes, inspect machinery and monitor product and materials handling.

The majority of product faults found in foodstuffs may be classified as 'visible' or 'non-visible'. Visible faults (those that can be detected by the human eye) can be sensed by normal video cameras. Many non-visible faults however, may be analysed in the same way as visible ones, given sensing methods which respond to electromagnetic wavelengths beyond that of human vision. It is therefore possible for one image processing work station to control and process different types of images, e.g. radiographic, thermal, ultraviolet. Figure 4.18 shows the relationship between the two classes of faults and how they are inspected.

Visual inspection systems are still relatively expensive, and so there is a tendency to inspect only at the end of the product line. This at least ensures that:

- The dimensions of the product are within an acceptable range.
- The final appearance is acceptable.

In the food industry, there is a trend towards products of high added-value. It is therefore logical that in the future, such products are inspected at every stage of manufacture. This would ensure that further value is not added to products that are already faulty in some way.

4.4.2 Shape and size

The size of the product is an important parameter to measure. On many food production lines, jamming of packaging machines due to oversized products is one of the major problems (Davies, 1990). The two-dimensional size of a product is a relatively straightforward parameter to measure if the edges of the product are clearly defined against the

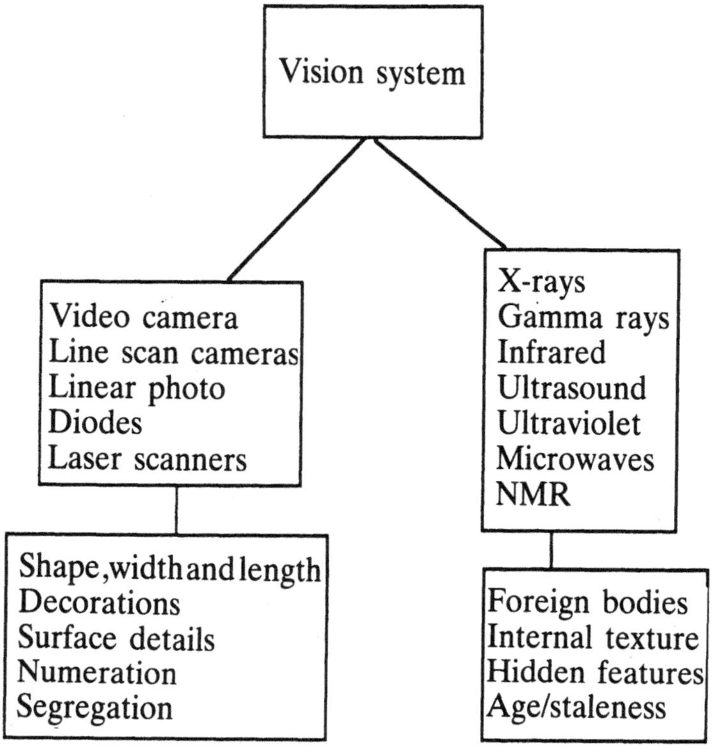

Figure 4.18 Tree diagram showing imaging methods for various product faults.

background (Figure 4.19). If this is not the case, it may be necessary to use 3-D techniques with structured lighting or laser lines (Marshall and Martin, 1992). Using such methods, it is also possible to determine the height

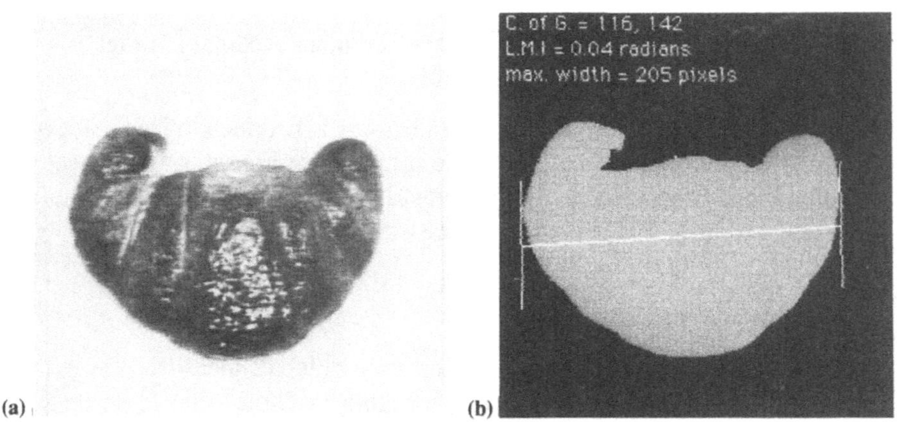

Figure 4.19 Measurements using machine vision: (a) original image of croissant, (b) after thresholding, the centroid, the angle of least moment of inertia and the maximum object width are found.

information and hence the volume of the product.

Visual measurement of size and shape can also be used in the grading and separation of products. For example eggs, potatoes, chickens, and so on. This task, as with many others in the food industry, is more involved than first appears. There are three main difficulties which have to be addressed before any grading can take place: speed requirements (to meet throughput), orientation of object being inspected, and separation of objects from their neighbours (often 'touching' and sometimes occluding one another). A method for overcoming some of these problems and the associated hardware architecture required are given by Marchant *et al.* (1990).

There are many ways to analyse and quantify the 2-D shape of a product, depending on the application. It is however, desirable that these methods are invariant to changes in position, scale and orientation. Some of the most common ones are:

- Shape factor, perimeter^2/area
- Number and areas of holes present (Euler number)
- Analysis of bays formed by the convex hull of the object (see Figure 4.20)
- Moments about the centroid
- Image morphology (Whelan and Batchelor, 1991)

4.4.3 Region labelling and object segmentation

Many food products contain separate distinct features, for example, the items in a complete set meal as provided by airlines, or the different components on a pizza (tomatoes, olives, peppers, etc). A region-labelling

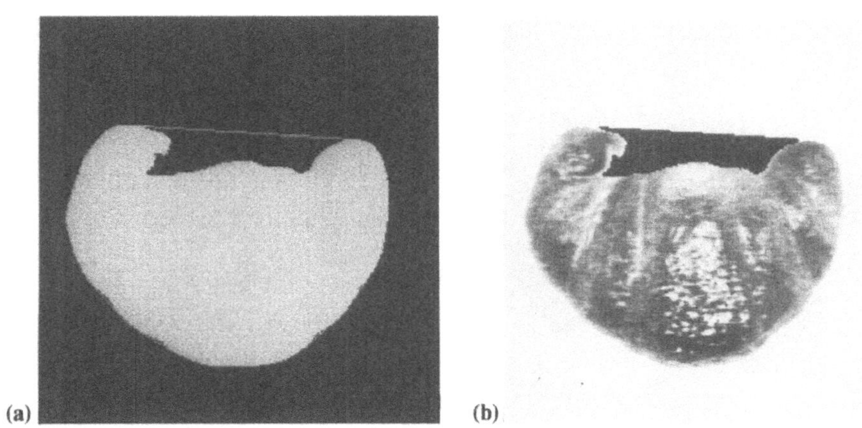

Figure 4.20 Using the convex hull of an image for shape analysis: (a) an 'elastic band' is wrapped around the image (convex hull), an exclusive OR operation reveals the 'bay' formed, (b) bay superimposed on original image.

operator will shade the objects according to the order in which they are found during a conventional raster scan of the input image (Figure 4.21). Alternatively, the objects could be shaded according to their areas, i.e. the biggest one becomes the brightest (or darkest). This is an extremely useful technique, since it allows objects to be separated and analysed one at a time. Small objects can be eliminated from an image if desired. It could also be used to simply count the number of distinct objects in the image.

In many applications, it is necessary to isolate groups of similar objects in an image for analysis. In such cases, the segmentation process would rely on the structural and colour properties of these objects. Figure 4.11 illustrated how certain features such as the cherries and the crimped pastry ring in a packet of bakewell cakes could be enhanced by using their colour attributes. Figure 4.22 shows how high level processing can then segment these components for analysis. Figure 4.22(a) is a dichromatic representation of these features. The pastry rings are extracted (Figure 4.22(b)) and inspected for width, continuity and concentricity. The cherries can also be segmented and analysed for their position on the cake and for size. Figure 4.22(c) is an image showing the cherries that are considered acceptable, note that the top left cherry is rejected due to the fact that it lies partly on the pastry. Finally, the centroids of the cherries are found (Figure 4.22d), reducing the complex original image into a small set of numbers.

4.4.4 Appearance of product

The appearance of a food product is often very difficult to quantify. For example, what constitutes a good loaf shape? And when does the appearance of a pizza become unattractive? A food product, unlike a precision-made engine component, is subject to natural variations inherent in its making. A human inspector would make his or her judgement on the basis of a complex ruleset, learned from extensive prior knowledge and

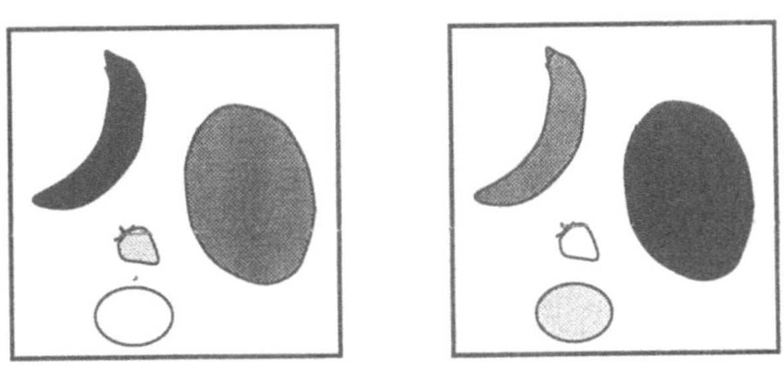

(a) (b)

Figure 4.21 Shading blobs in a binary image: (a) according to order found during a raster scan (left to right, top to bottom), (b) according to areas.

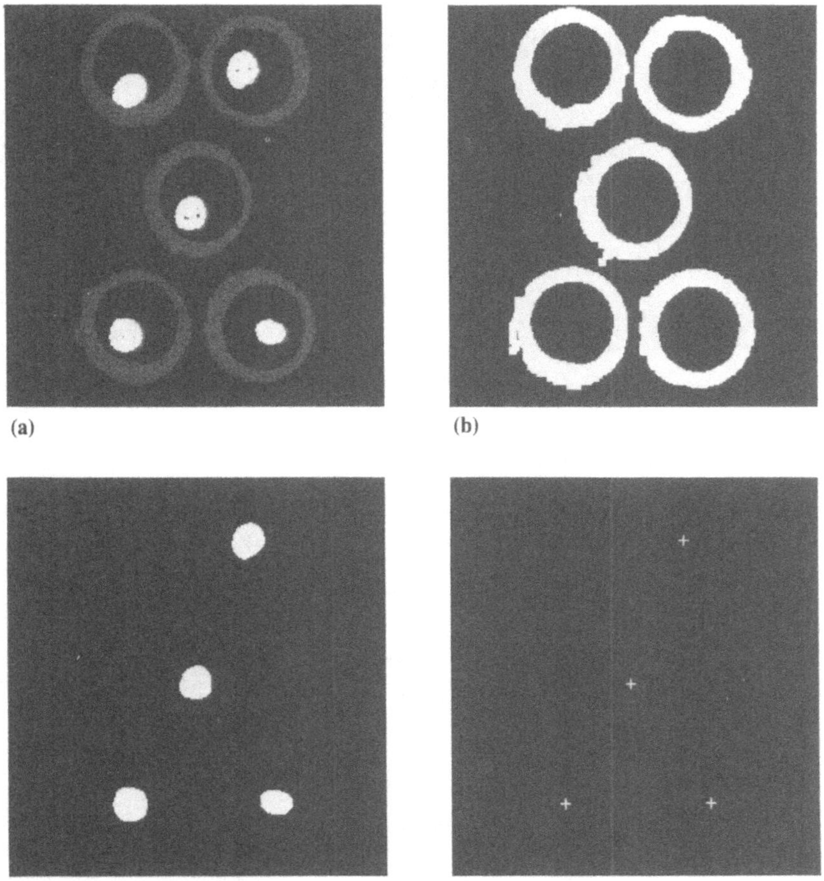

(a) (b) (c) (d)

Figure 4.22 Segmentation of similar components for analysis: (a) dichromatic representation of components, (b) crimped pastry ring, (c) isolating acceptable cherries, (d) centroids of cherries.

experience. Computer learning methods do exist, but it would be foolish to think that even the most powerful computer system could come close to matching the qualitative and interpretative powers of the human mind. By extracting and analysing certain individual features, it is possible to form a quantitative description of the appearance of a food product in a limited way.

The appearance of a decorated cake can be analysed in such a way (Figure 4.23). The zig-zag pattern on the cake is first enhanced to highlight it from the icing. The image is then processed so that only the pattern remains, as a series of connecting white pixels. A set of rules is applied to the image ensuring no obvious faults are present, e.g. breaks in the pattern. Suitable parameters are then extracted from this image and

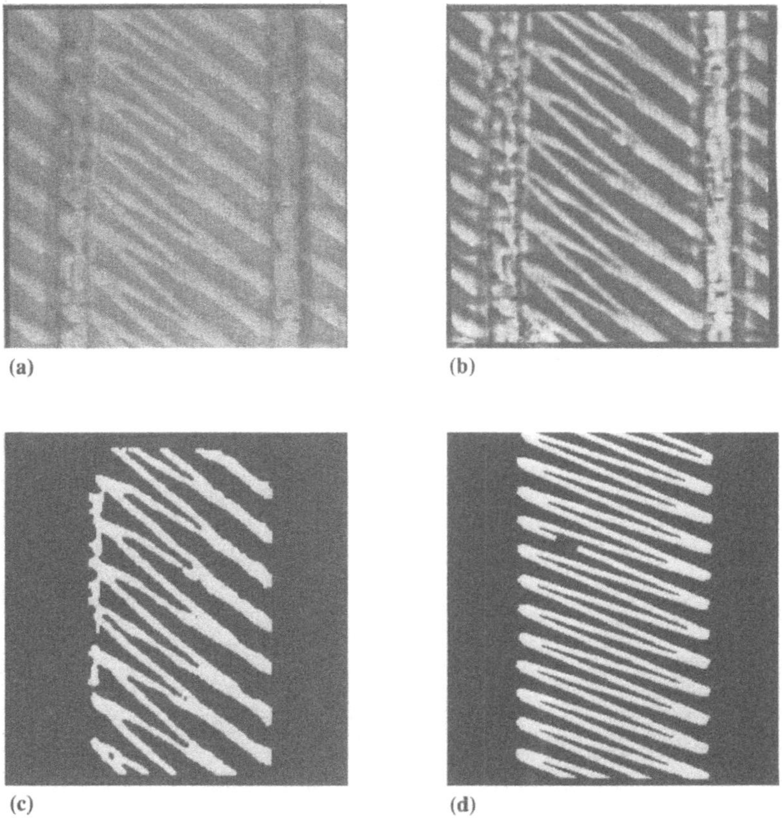

Figure 4.23 Inspecting decorations on a cake: (a) original image, (b) enhanced image, (c) isolating pattern decoration, (d) pattern with a break.

compared with a previously learned knowledge-base, in order to check for the integrity and conformity of the pattern. The appearance of the cake can then be classified as acceptable or otherwise.

Obvious surface defects are much easier to inspect and are based on detecting an abrupt change in the characteristics of the surface. Many techniques have been developed to do this (Gonzales and Wintz, 1982). They often involve deriving certain measures from a sequence of pixels along the surface and noting a change or deviation in the sequence. A simpler approach and often more robust, is to compare a series of images produced using simple image processing operations, and look for any changes. An example of this is crack detection (Figure 4.24).

Figure 4.24 shows a baked product which has a crack on its surface. A largest neighbour averaging technique is applied to produce a new image. This is essentially a morphological operator which 'fills in' the crack. When this image is subtracted from the original image, the crack becomes

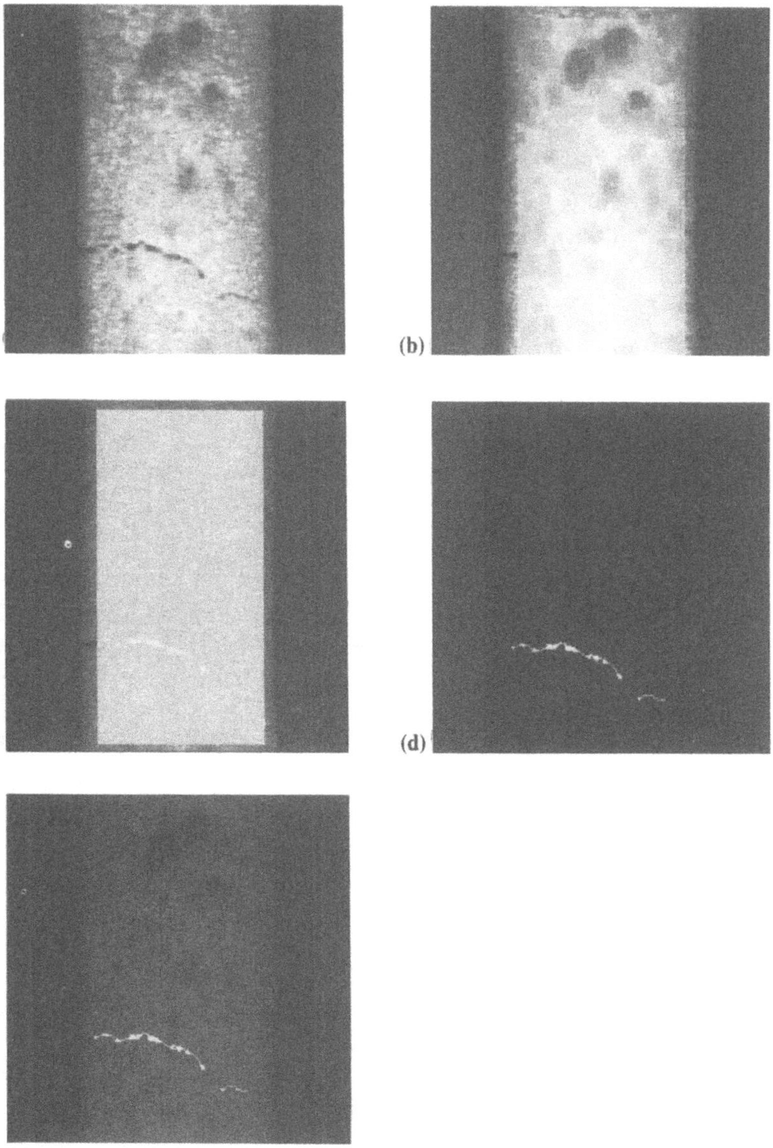

Figure 4.24 Crack detection: (a) original image, (b) result of morphological operator to remove crack, (c) subtracting ((a) from (b)), (d) crack is found, (e) crack superimposed on original image.

obvious and can be highlighted. This method of crack detection is invariant to gradual changes in intensities across the product, and would be ideal for the inspection of cracks in eggs.

Simple correlations of grey scale intensities can give us much information about a product based on surface details. The ripeness of tomatoes for example (see section 4.2), and the level of bake of a cake, the darker it is, the more baked it is.

4.4.5 Texture

Sometimes it is easier to describe what we are looking at by means of its texture, rather than its shape. The concept of texture, however, is a rather difficult one to pin down exactly, but 'something comprising one or more simple elements repeated in a periodic or random manner' comes close. It is clear that many food products fulfil this criteria.

There are two main approaches for recognising and classifying textures. The first method, the structural method, considers the texture to be composed of primitive elements arranged in a repeated manner. The second method, the statistical method, describes a texture in terms of the distribution of pixel values in the image, for example in terms of the number of pixels of each intensity, or the spatial relationships between pixels of differing intensities, or both.

In food products, the former approach is better suited to textures containing relatively large primitives which can be readily identified, and to textures where the pattern is regular. For example, tea leaves and decorations on cakes. The statistical approach is more useful when looking at random textures where the primitives are variable or otherwise difficult to identify. A good example of this is the texture of baked products such as bread. Further discussion of both methods is well documented in Ballard and Brown (1982).

4.4.6 Non-visual imaging techniques

Experiments have shown that where faults cannot be detected by normal visual means, the use of other sensing methods yielded extremely encouraging results (Chan *et al.*, 1990; MacAndrew and Harris, 1991). X-ray imaging for example, was able to detect small foreign bodies if they were of higher density than the food sample (Dykes, 1985; Chan, 1988). In thin uniform structures, low density foreign bodies such as air voids were also visible. Figures 4.25 and 4.26 illustrate how X-ray images can be processed to identify the presence of foreign bodies. Figure 4.25(a) shows a glass jar containing jam contaminated with small fragments of glass foreign bodies. Figure 4.25(b) is the result after the application of an image processing program, the glass fragments are detected and highlighted. Figure 4.26 shows a similar application for inspecting wheat grains for stones.

Table 4.1 is a summary of the detectability of various foreign bodies in different food products, based on experimental results (Chan *et al.*, 1990).

X-ray imaging is also useful in the analysis of the internal texture of the food product (Chan, 1988). For example, an unsliced loaf of bread may be

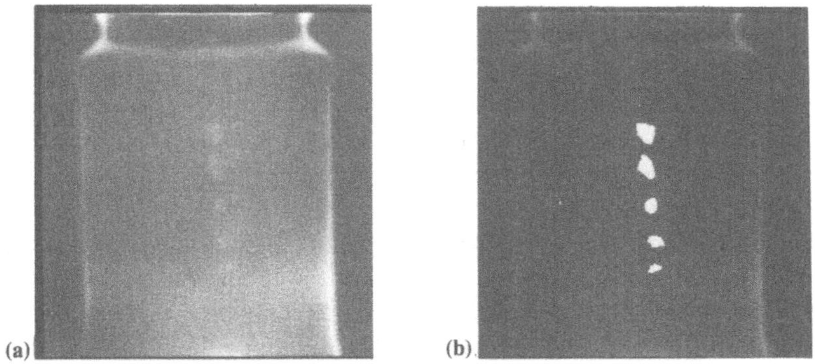

Figure 4.25 Inspection for glass foreign bodies in jar of jam: (a) original image, (b) glass fragments highlighted after processing image.

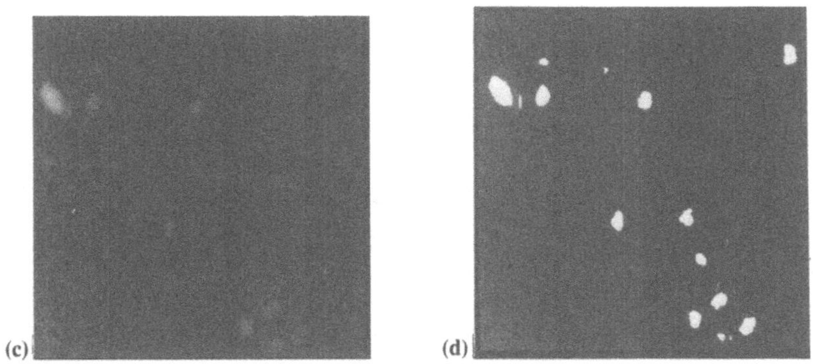

Figure 4.26 Inspection of wheat grains for stones: (a) original image, (b) stones highlighted after processing image.

Table 4.1 Summary of detectability of foreign bodies in foods using X-rays

	Glass	Wire	Bone	Beetle	Perspex	Stones	Hair
Cooked meat	5	5	4	2	2	3	0
Plain flour	5	5	5	4	2	4	1
Coleslaw in pot	5	5	4	2	1	3	0
Jam in glass jar	3	3	2	1	0	1	0
Bread in cellophane	5	5	5	2	1	4	0
Muesli	5	5	2	0	0	2	0

Score: 5 = very detectable, 0 = not detectable.

inspected to ensure it has risen normally, containing a correct distribution of void sizes. It can ensure that the correct level of ingredients such as granary wheat in bread, or raisins in fruit cakes, are present. It can also be used to inspect for mixing homogeneity, ensuring consistent quality.

Although ultraviolet light does not penetrate materials like X-rays, it is nevertheless useful in the inspection of certain materials using ultraviolet luminescence. Eggs, milk and fish all show marked changes in fluorescence with age (freshness). Fresh milk for example, has a yellow fluorescence whereas soured milk appears white to grey violet. Another example is the inspection of peas for mould and bacteria. The normal fluorescence of peas is replaced by black or bright yellow if bacteria or mould, produced by storing in a damp place, are present (White, 1970; Perry, 1990).

Nuclear magnetic resonance could also play a part in future inspection systems. Recent research has shown that it is possible to reveal the texture and composition of foods by using NMR. It shows for example, whether a soggy pork pie results from water or from fat in the crust. It can also reveal bruising of fruits and the temperature distribution in food products. The drawbacks at present are speed and cost: approximately £250 000 (1992) for a scanner big enough to accommodate a chicken, taking anything up to 30 minutes for a high resolution scan.

Table 4.2 is a tabulated summary of the different techniques that may be used in the inspection of food products.

4.4.7 Object segmentation using colour

Colour is, of course, of major importance to many food products and their packaging materials. The proper distribution of tomatoes and peppers on pizzas is important, as is the visual appearance of the decoration patterns on cakes and the location of the cherries on Bakewell tarts. Ensuring that boxes of foil-wrapped sweets contain a correct distribution of various types is also important for product quality. There are, of course, many other potential applications for colour recognition in the food industry.

There are, in addition, many possible approaches to colour recognition. The one that we are about to describe was pioneered by Image Inspection, Epsom, Surrey, UK. It is capable of only coarse colour discrimination and it is not, for example, possible to use this technique to examine bread or pastry products for the degree of baking. Nevertheless, there are many instances where it is important to be able to decide whether a part of a scene is red, or green, or yellow, or to estimate the proportions of these colours.

The RGB (signifying red, green and blue) outputs from a colour camera are each digitised, with an precision of 6 bits. The resulting 18 bits are then used as the (address) inputs to a random access memory (RAM), of capacity 2^{18} bytes (= 256 bytes). The output consists of a set of 8 parallel signal lines. The RAM acts as a look-up table, (LUT). By loading appropriate values into the LUT, it is possible to recognise a wide range of colours. Various ingenious procedures have been devised for selecting suitable values for storage in the LUT RAM and it is possible, for example, to learn the range of variation of the colours associated with a certain type of product. The recognition process is very fast; it is possible to perform

Table 4.2 Basic comparison between imaging techniques

	Ultrasound	X-rays	Gamma-rays	Ultraviolet	Infrared
Application to foodstuff	Packaged foodstuff with no air content, and high water content	Low density materials with low water content; foreign body detection and texture analysis	Similar to X-rays, but denser materials may be inspected	Fluorescent analysis of certain foodstuffs	Thermal imaging of foodstuffs
Speed	Real time imaging possible	Real time imaging possible	Real time possible	Real time imaging possible	Inspection rate approx. 5 samples per sec.
Safety requirements	None	Considerable	Considerable	At high intensities, eye damage or skin cancer may occur	None
Additional system requirements	Temperature must remain constant, and coupling medium is required	Image intensifier converts to visible spectrum. System must be caged in lead	Similar to X-rays	Effect of ultraviolet on the foodstuff must be assessed	Environmental temp must remain relatively constant
Flexibility	Low flexibility, transducer must be made specifically for material to be imaged	System easily adapted to different imaging applications	Only distance or type of source may be altered to change imaging properties	Suitable only for certain materials	Good flexibility, and recalibration is simple
Other advantages and disadvantages	Gel or water interface between transducer required. Transducer not well suited to industrial environment	Safe when switched off	Source cannot be switched off, creating storage and transport complications. Beam cannot be directed	Colour vision necessary for fluorescent analysis	Camera examined limited to 0.5 degrees sensitivity

Table 4.2 Basic comparison between imaging techniques (*cont'd*)

	Ultrasound	X-rays	Gamma-rays	Ultraviolet	Infrared
Factors affecting image quality	Propagation velocity of material, differences in accoustic impedance, intensity reflection coefficient and attenuation	Amperage (image density), Voltage (penetration & image contrast). Adsorption coefficient of sample. Geometry of set-up also important	Type and distance of source from object	Amount of ultraviolet and visible light entering system	Size of differences in temperature distribution
Ease of image processing	Not encouraging, but probably better if specific transducer was designed resulting in less spurious data	Generally, images of high contrast, and may be easily analysed	Similar to X-ray images, but less contrast	Colour processing may be required. Maybe used as form of contrast enhancement	Pre-processing required

colour recognition in real time on a video signal. Moreover, the cost of the equipment needed to do this is quite modest. Self-adaptive learning procedures can be devised, so that it would be possible, for example, to update the machine, to accommodate new or time-varying products. It is a straightforward matter to combine colour recognition with other image data, describing for example, the shape or texture of the product.

To illustrate the possibilities of colour recognition, consider Figure 4.27, where it will be seen that it is possible to discriminate between the colours of certain fruit and vegetables. The authors have studied various processed food products and their packaging materials and can report that favourable results have been obtained.

It must be emphasised that fine colour discrimination requires a different approach, with a better quality sensor. The outstanding example where colour information has been used is in the 'Automated meat piece grading system' (Holmes, Newman and Associates, 1992). This instrument is able to measure the fat/lean ratio of various types of meat. We have already hinted at the need for a system that can monitor the degree of baking. This also requires a high level of precision.

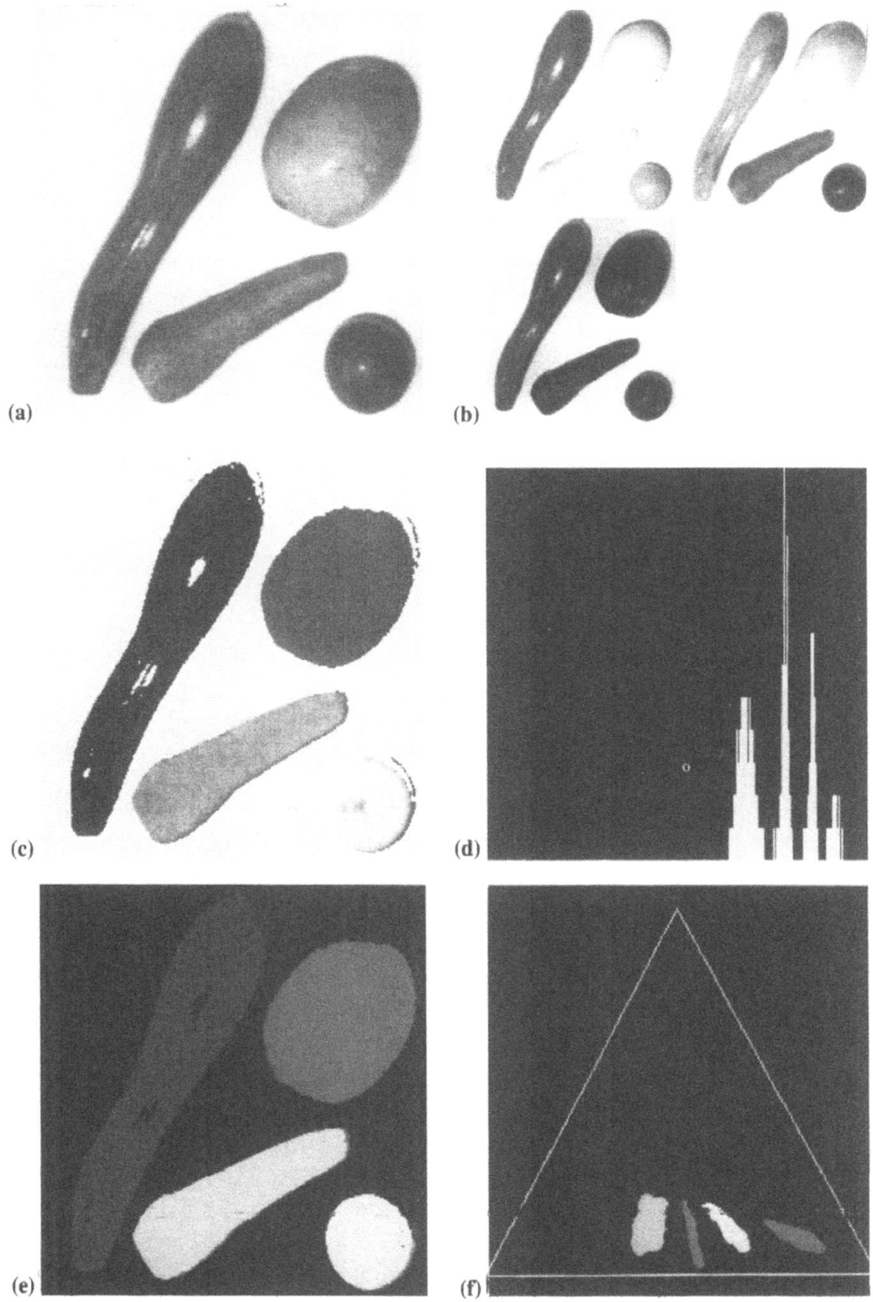

Figure 4.27 Object recognition using colour: (a) grey scale image consists of courgette, lemon, carrot and tomato, (b) RGB components of image (top left, red; top right, green; bottom, blue), (c) 18-bits colour image transformed in 8-bit image, (d) intensity histogram showing four distinct peaks (left to right: green, yellow, orange, red), (e) four-grey-level representation of objects found, (f) colour triangle showing four major colours in scene being viewed (left to right: lemon, carrot and tomato).

The authors believe that there remains an outstanding need for systems that can perform coarse colour discrimination. While certain applications clearly require fine discrimination, there are many others where the ability to recognise 'yellow' or 'green', or 'red' is important.

4.4.8 Process monitoring and control

So far we have largely discussed the use of machine vision for product inspection. This application of machine vision is one with which the food industry can most easily identify. Vision systems can also have great potential in other areas, such as: monitoring individual processes to ensure that the correct action is taken (cutters, depositors, mixers, etc); controlling electromechanical devices for manufacture and assembly (lights, robot arms, etc); and checking the general factory environment. There are endless possibilities, some of them are listed as follows:

- Checking that gangways are clear
- Ensure that fire doors are shut
- Assess stacking areas for capacity and safety
- Check equipment for cleanliness
- Monitor conveyor belts for jams
- Detect unauthorised tempering of machinery and products
- Check for water and oil spillages

Packaged meals, such as those supplied to airlines, present both an inspection and an assembly problem. A vision system must first locate the food items and instruct a robot or similar mechanism, to place individual items onto the tray. The system then has to inspect the food tray for faults and appearance, for example the fruit salad must contain a correct distribution of various fruits. The vision system may also inspect the food items before the mechanical handling, to avoid wasted effort and damage to the robot. Here, AVI and robot vision (RV) work together to optimise the usage of a machine vision system. This blurring of AVI and RV is likely to become more pronounced as vision systems are increasingly used for complex and highly variable products such as those found in the food industry.

4.4.9 High level techniques for machine vision

As machine vision tasks become more complex and variable, there is a need for adaptive systems that possesses high levels of machine intelligence. In designing a visual inspection system, there are many factors, often conflicting, to be considered. Invariably there will be a trade-off between flexibility, inspection robustness, speed and cost. In the following example, we can demonstrate how a vision system can overcome some of these limitations by integrating it into an expert system with an expandable knowledge base. Such a method enables the system to adapt to limited changes in the inspected product without the need for an experienced vision programmer.

An expert system has been written in the AI language, Prolog, which enables a non-expert user to specify the method of manufacture and visual appearance of small decorated cakes (Chan *et al.*, 1991). The expert system is designed to be ergonomically acceptable to non-expert users. It formulates the inspection procedures necessary by relating the method of manufacture to the appearance of a product (Figure 4.28). From this, the image processing operations are generated in the form of a Prolog+ program. Such a method enables the system to inspect new and varied products without sacrificing the complexity and robustness of inspection. Other advantages include:

- The automated learning procedures are applied to a narrower range of product types, since the expert system will have already acquired knowledge from the user about the manufacturing and appearance of the product. Hence inspection parameters can be learned with much greater reliability and accuracy.
- The system would be much more flexible to new features and variations in new products.

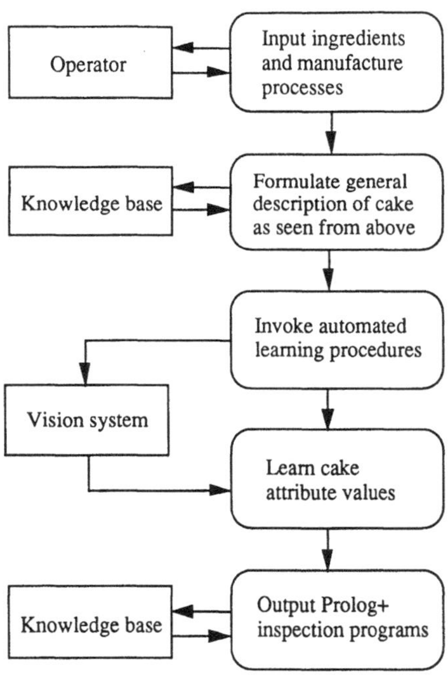

Figure 4.28 Schematic diagram summarising expert system structure for learning and inspection of decorated cakes.

- Variability within a given product is more easily dealt with, since learning parameters satisfy a set of nebulous rules, rather than intolerant numerical specifications.
- The human interaction creates a degree of confidence in the system, since the user can be assured that the machine does not perform any unnecessary tasks or go down a 'blind alley'.
- Any limitations of the inspection routines can be quickly realised and reported.

Applying knowledge based expert systems to meat grading has also been documented (Chen, 1989; Holmes Newman and Associates, 1992), amongst others. The use of high level techniques such as expert systems, natural language, speech synthesis, speech recognition, declarative programming and neural networks will increasing play a major role in future vision systems. These techniques not only enable target systems to be more 'intelligent', they can be extremely useful tools to minimise the effort required to design a machine vision system (Batchelor, 1990).

4.4.10 Future applications of machine vision systems

As we have seen in our brief consideration of machine vision techniques in this chapter, many faults and processes may be inspected with such techniques. However, the solutions developed for industrial use have so far been somewhat *ad hoc* in nature and thus are very specific in their application. In many ways industry has always favoured this approach (Freeman, 1988), since solutions are dealt with case-by-case and tend to be simpler and more compact. For general purpose quality assurance and process monitoring, particularly in the food industry, such machines will not be cost effective in the future due to their high development costs. The potential exists for general purpose machine vision systems to be constructed, capable of performing a variety of inspection and monitoring tasks. Indeed, many such prototype systems are being built or investigated at the present time (Rummel, 1988; Batchelor, 1990). A realistic general purpose vision system would include a variety of different imaging methods and a flexible lighting arrangement as well as robot arms for pick and place tasks and non-visual sensors for more specific tasks. The food industry is one sector that could reap huge benefits from machine vision techniques. Indeed, inspection and monitoring of such products and their production processes will no doubt provide many interesting challenges for machine vision systems for many years to come.

References

Ballard, D. H. and Brown. C. M. (1982) *Computer Vision*, Prentice-Hall, New Jersey.

Batchelor, B. G. (1979) Interactive image analysis as a prototyping tool for industrial inspection, *Computers and Digital Techniques*, **2**, 61–69.

Batchelor, B. G. (1989) A Prolog lighting advisor, *Proc. Internat. SPIE Conf. Intelligent Robots and Computer Vision VIII: Systems and Applications*, Philadelphia, pp. 168–173.
Batchelor, B. G. (1990) Tools for designing industrial vision systems, *Proc. Internat. SPIE Conf. Machine Vision Systems Integration*, paper CR 36–13, Boston.
Batchelor, B. G. (1991) *Intelligent Image Processing in Prolog*, Springer Verlag, London.
Batchelor, B. G., Hill, D. A. and Hodgson, D. C. (1985) *Automated Visual Inspection*, IFS Publications, Bedford, England.
Batchelor, B. G., Harris I. P., Waltz and Snyder, M. A. (1988) Applications of the ProVision language in robot vision. *Proc. SPIE* Conf., Cambridge, MA, USA.
Bowman, C. C. and Batchelor, B. G. (1986) Automated visual inspection. In *Non-destructive Testing* (ed. Sharpe, R. S.) vol. 8, pp. 361–444.
Castleman, K. R. (1978) *Digital Image Processing*, Prentice-Hall, Englewood Cliffs, USA.
Chan, J. P. (1988) *Automated X-ray Inspection of Foodstuff*, MSc thesis, UWCC, Cardiff, UK.
Chan, J. P. and Batchelor, B. G. (1990), Integrating vision and AI in an image processing workstation. *Proc. SPIE, Internat. Soc. Optical Eng., Machine Vision Systems Integration in Industry*, Boston, MA, USA, vol.1386.
Chan, J. P., Batchelor, B. G., Harris, I. P. and Perry, S. J. (1990). Intelligent visual inspection of food products. *Proc. SPIE, Internat. Soc. Optical Eng., Machine Vision Systems Integration in Industry*, Boston, vol.1386, pp. 171–179.
Chan, J. P., Batchelor, B. G., Broderick, S. R. and Weeks, A. W. (1991) Expert system to aid learning of new products. *Internat. Soc. Optical Eng., Machine Vision Systems: Integration and Applications*, Boston.
Chen, Y.-R. (1989) Applying knowledge-based expert system to meat grading. *Proc. Annual AI Systems in Government Conference* (IEEE Cat. No.89CH2715–1) IEEE, Washington, pp. 120–123.
Davies, E. R. (1990) *Machine Vision: Theory, Algorithms, Practicalities*. Academic Press, London.
Dykes, G.W. (1985) Automated inspection of food jars for glass fragments. *VISION '85 Conf. Proc.* Machine Vision Assoc. SME, USA, pp. 6/21–6/30.
Freeman, H. (1988) Is industry ready for machine vision. In *Machine Vision for Inspection and Measurement*, Academic Press, San Diego, USA.
Gonzalez, R. C. and Wintz, P. (1982) *Digital Image Processing*. Addison Wesley, Cambridge, MA.
Holmes, Newman and Associates (1992) *Automated Meat Piece Grading System*, Oakhampton, Devon. Patent pending.
Lightbody, M. S. (1989) New technological approaches to reducing uniformity in processed foods. *Br. Food J.*, **110**.
MacAndrew, A. and Harris, C. (1991) Sensors detect food contamination. *Sensor Rev.*, **11** pp.23–25.
Marchant, J.A., Onyango, C.M. and Street, M.J. (1990) Computer vision for potato inspection without singulation. *Computers and Electronics in Agriculture*, **4** (3), 235–244.
Marshall, A. D. and Martin, R. R. (1992) *Computer Vision, Models and Inspection*. World Scientific, Singapore.
Perry, S. J. (1990) *An Analysis of Non-visual Imaging for Machine Vision*, MSc thesis, UWCC, Cardiff, UK.
Reid, G., Rixon, R. and Messer, H. (1984) Absolute and comparative measurements of three dimensional shape by phase measuring moire topography, *Optics and Laser Technol.*, **16** 315–319.
Rummel, P. (1988) Combining workpiece recognition and inspection. In *Machine Vision for Inspection and Measurement*. Academic Press, San Diego, USA.
Whelan, P. (1991) *Dublin City University Course notes- Machine Vision*, Dublin City University, Ireland.
Whelan, P. and Batchelor, B. G. (1991) Automatic packing of arbitrary shapes. *Proc. SPIE, Machine Vision Systems: Integration and Applications*, Boston.
White, C. E. (1970) *Fluorescence Analysis*. M. Dekker, UK.

Bibliography

Adaway, B. (1989) Fruitful inspection (machine vision). *Image Processing*, **1** (2), 18.

Aldersey-Williams, H. (1985) Machine vision finds new applications. *Digital Design*, **15** (6), 40–46.

Batchelor, B. G. (1974) *Practical Approach to Pattern Classification*. Plenum Press, London.

Batchelor, B. G. (1987) An interactive AI language for image processing and robotics. *Proc. SPIE, Intelligent Robots and Computer Vision: Sixth in a Series*, Cambridge, MA, USA.

Batchelor, B.G. (1989) Integrating vision and AI for industrial applications, *Proc. SPIE, Internat. Soc. Optical Eng., Intelligent Robots and Computer Vision VIII: Systems and Applications*. Philadelphia, vol.1193, pp.168–173.

Cormack, R. H. and Brown, C. (1989). Aluminum can metrology using sub-pixel image measurement. *VISION '89 Conf. Proc.* Soc. Manuf. Eng, Dearborn, MI, USA, pp.8/47–8/60.

Davies, E. R. (1986). Corner detection using the generalised Hough transform. *Proc. Second Internat. Conf. Image Processing and its Applications* (Conf. Publ. No.265), IEE, London, pp.175–179.

Dennis, T.J. and Clark, L.J. (1986). Real time detection of spot-type defects. *Proc. SPIE, Internat. Soc. Optical Eng, Architectures and Algorithms for Digital Image Processing*, vol.596, pp.178–183.

Dowell, F.E. (1990) An intelligent automated system for determining peanut quality. *Proc, IROS 90. IEEE Internat. Workshop Intelligent Robots and Systems '90. Towards a New Frontier of Applications* vol.1, pp.237–41.

Dyche, R.K. (1991) Inex-100 per cent online visual inspection of consumer products. *Sensor Rev.*, **11** (4), 14–16.

Edmonds, J.M. and Davis, E.R. (1991) High-speed processor for realtime visual inspection. *Microprocessors and Microsystems*, **15** (no.1), 11–19.

Gagliardi, G., Hatch, G.F. and Sarkar, N. (1985) Machine vision applications in the food industry. *VISION '85 Conf. Proc.*, Machine Vision Assoc. SME, USA, pp. 6/40–6/54.

Gerber, G.L., Holmes, Q.A. and Calhan, R. (1985) Industrial machine vision with X-ray sensor for online food processing inspection. *Proc., SENSORS '85 Conf.* Comput. and Autom. Syst. Assoc. of SME, USA, pp.MS85–1009/1–17.

Goodenough, D.G. and Fung, K. (1991) Knowledge-based methods for temporal and spatial data fusion. *Canadian Conf. GIS, Proc.* Ottawa, Ontario, Canada, pp.994–995.

Hamza, M.H. (ed.) (1989) *Proc. IASTED Internat. Symp., Applied Informatics—AI'89.* ACTA Press, Anaheim, CA, USA.

Hodgson, R.M. and McNeill, S.J. (1986) The design of image processing systems for real-time inspection applications. *Proc., Second Internat. Conf. Image Processing and its Applications* (Conf. Publ. No.265) IEE, UK, pp.126–129.

Keller, J.M. (1986) Color image analysis of food, *Proc. IEEE Computer Soc. Conf. Computer Vision and Pattern Recognition*, Florida.

Khodabandehloo, K. (1989) Getting down to the bare bones (robotic meat cutting). *Industrial Robot*, **16** (3), 160–5.

Lord, J. (1989) Graphics rules OK. Conf. Proc. *Seventh Australasian Conf. Computer Graphics*—Ausgraph 89. Australasian Comput. Graphics Assoc, Parkville, Victoria, Australia, pp.43–50.

MacDougal, D.B. (1990) Colour vision and appearance measurement, In *Sensory Analysis of Food*. (ed. Piggott, J.) Elsevier Applied Science.

Marchant, J.A. (1990) A mechatronic approach to produce grading. *Proc. Institution of Mechanical Eng., Internat. Conf., Mechatronics: Designing Intelligent Machines*, (IMechE 1990–8), Cambridge, UK, pp.159–64.

Marchant, J.A. and Cowlin, S. (1990) Parallel processing for produce inspection. Proc. *IEE colloquium on Parallel processing: Industrial and Scientific Applications* (Digest No.122) IEE, London, pp.8/1–4.

McClelland, S. (1987) A sweet sense of success for PA technology (food processing applications of vision and robots). *Sensor Rev.*, **7**, 179–82.

Mizier, M.-O., Nguyen, V. and Piquard, J.F. (1988) Foreign bodies in the agricultural and food industries: when X-rays ferret out the unwelcome. *Mesures*, **53** (11), 59–61.

Naylor, P. (1988) Machine vision techniques—process control and product quality assurance. *Proc., FACTORY 2000: Integrating Information and Material Flow* (Publ. No.80) IERE, London, pp. 127–32.

Neuman, M.R., Shwedyk, E. and Bushuk, W. (1989) A PC-based colour image processing system for wheat grain grading. *Proc., Third Internat. Conf Image Processing and its Applications* (Conf. Publ. No.307), IEE, London, pp.242–246.

Optics in Agriculture (1990) *Proc. SPIE, Internat. Soc. Optical Eng*, vol.1379.

Richardson, A.J. and Blazquez, C.H. (1989) Estimating citrus grove production using image processing techniques. *J. Imaging Technol.*, **15** (6), 272–276.

Rion, C., Evrat, C., Roignot, D. and Tosser, A. (1989) Real-time automatic inspection of moving corn ears. *Proc. IASTED Internat. Symp. Modelling Identification and Control—MIC '88*, ACTA Press, Anaheim, CA, USA, pp. 326–331.

Sacks, T. (1988) Industry zooms in on machine vision. *Electrical Rev.*, **221** (13), 16–17.

Sevila, F. and Baylou, P. (1991) The Principles of Robotics in Agriculture and Horticulture. In *Progress in Agricultural Physics and Engineering* (ed. Mathew, J.). CAB International, Wallingford.

Shinoda, T., Ohnaka, S., Kawada, M., Motonaga, H. and Koyanagi, K. (1990) Automatic inspection systems of empty bottles, and foreign substances by CCD cameras and image processing technology. *NEC Tech. J.*, **43** (11) 64–70.

Tachi, S. (1991) Systemization and its applications: TELE-EXISTENCE. *J. Inst. Television Eng. Japan*, **45** (4) 480–487.

Tailor, A., Corr, D.G., Cosoli, P., Cross, A., Hogg, D.C., Lawrence, D.H., Mason, D.C., Petrou, M. and Vango, R.D. (1987) A system for knowledge-based segmentation of remotely-sensed images. *Proc., IGARSS '87. Remote Sensing: Understanding the Earth as a System*. IEEE, New York, vol.1, pp. 111–116.

Taylor, R.W. and Rehkugler, G.E. (1985) Development of a system for automated detection of apple bruises. *VISION '85 Conf. Proc.*. Machine Vision Assoc. SME, USA, pp.6/55–6/65.

Tillet, R.D. and Batchelor, B. G. (1991) An algorithm for locating mushrooms in a growing bed. *Computer and Electronics in Agriculture*, **6**, 191–200.

Tucker, J.W. (1989) Inside beverage can inspection. An application from start to finish. *VISION '89 Conf. Proc.*. Soc. Manuf. Eng, Dearborn, MI, USA, pp. 1/1–1/12.

Vasilash, G.S. (1989) A new vision for cost and quality. *Production*, **101** (6), 70–74.

Wagner, G.G. (1987) Combining X-ray imaging and machine vision. *Proc. SPIE, Internat. Soc. Optical Eng.*, vol.850, p.45–53.

Watanabe, H. (1986) Application of general-purpose visual recognition system to food industries. *Robot*, **54**, 32–39.

Wilkie, F. (1986) Industry opens its eyes to machine vision. *Process Eng.*, **67** (1), 41–43.

Wu, P.S. (1988) Attribute identification to agricultural product inspection. Robots: coming of age. *Proc. Internat. Symp. Exposition on Robots* (ed. Jarvis, R.A.). IFS Publications, Kempston, UK, pp. 440–450.

Young, G.L. (1988) High speed, high resolution inspection systems for plastic microwaveable trays. *Proc. Electronic Imaging '88: Internat. Electronic Imaging Exposition and Conf.*. Advance Printing of Paper Summaries vol.2. pp. 687–688.

5 Particle sizing

M. WEDD

5.1 Introduction

Since many foods, spices and flavourings are prepared or used in a finely divided powder form, it is hardly surprising that the study of particle size distribution has been important to the food processing industry. The appreciation of small, young, early peas as the most flavoursome, has been instilled in many people since childhood. Whilst the final appreciation of this vegetable depends upon many other factors, size is a key issue. Another catch phrase 'graded grains make finer flour' implies that both before and after milling, grain and particle size are considered. The direct sales of apples and eggs are classified by size and priced accordingly. This introduces the concept of a premium price for products of the 'correct size'. In general any quality product is a graded product, which fulfils a spectrum of requirements of the customer.

Modern food processing and supply, particularly of basic commodities like sugar, is applied to a market that has little or no natural growth. However, pressure on controlling production costs continues. The characterisation of particle size distribution is a crucial aspect of cost and quality control in many food processing and supply aspects, and the adage 'you cannot manage what you cannot measure' has been fully appreciated.

5.2 Particle sizing within the food processing industry

The dynamic range of particle sizing conducted within the food industry covers both extremes of size. It ranges from micelle and colloid structures in milks and flavourings as small as 15–20nm, up to tens of mm for nuts, dried fruit and seeds. No one method of particle size analysis has yet been devised to cover the whole of this vast range. It is relevant, however, briefly to look at the more common methods of particle sizing currently in use.

5.2.1 Sieving

Sieve analysis is one of the most widely used methods of particle sizing. It is comparatively inexpensive in capital terms to get started, but as it is a labour intensive operation, the cost per analysis can be high if a large number of size determinations is required on a regular basis. Woven wire sieves are produced which cover the approximate range of 20 μm to 125 mm. Micromesh sieves extend the range down to about 5 microns, with punched plate sieves extending the upper range.

Sieving is either conducted on dry powders or by wet sieving, when the particles are suspended in a liquid. The degree to which particles have a tendency to stick to each other determines which is the most appropriate method.

For materials that exhibit a tendency for the particles to adhere lightly to each other, either due to electrostatic or chemical bonding forces, there is always a tendency for small particles to attach to larger ones and thus deplete the mass of material in the finer fractions. Mechanical sieve shakers are usually operated for a finite time. Extending the run time increases the probability that the fine particles may be dislodged from their attachment to the larger ones. It has often been observed that the longer the sieving, the finer the result.

In addition, it is also possible for some particle attrition, due to mechanical action on the particles, to take place. Wet sieving, where appropriate, enables the increased dispersion power of liquids to be utilised, with or without additional wetting agents. For very fine micro-sieves and to a lesser extent for the coarse woven sieves, it is perfectly possible for fine particles to form bridges over the apertures and thus be held on a mesh much larger in aperture than the size of the individual grains. Tapping the dry sieve is often sufficient to overcome this problem. With wet sieves the use of additional clean dispersant flushed through, often suffices to remove this problem.

For irregular particles it should be borne in mind that the sieve discriminates on the second largest dimension of the particle. The result of the sieving process is reported as the weight of material retained on a specific size of sieve.

Consider a solid particle shaped like a beer can. The final sieve likely to hold a beer-can shaped particle is one whose aperture is just too small to allow passage of the diameter of the can. As the length of the beer-can shaped particle is about 2.5 times its diameter its weight will therefore be greater by the same ratio.

Sieving irregular-shaped particles has the tendency to bias the reporting of the distribution in favour of the smaller sizes, by the mechanism just described. In consequence, any mean values calculated from such a distribution are also biased towards the smaller in size.

5.2.2 Sedimentation

Particles suspended in a fluid and settling under gravity have been used to determine particle size since the early work of Andreasen in the late 1940s and early 1950s.

A particle in a liquid is acted upon by three forces. The mass of the particle with gravity produces a downwards force. The mass of liquid displaced by the particle with gravity form a buoyancy force upwards. The third force is the drag experienced by the particle as it moves through the fluid.

If a spherical particle has a greater density than that of the fluid, it will experience an acceleration towards the bottom of the container. This resultant force is opposed by the drag forces of friction in the fluid and when the opposing forces are equal, a terminal velocity is reached.

Provided the terminal velocity is small, i.e. creep flow, such that no turbulence is created by the passage of the particle through the fluid, then the conditions for Stokes' settling velocity for a sphere have been met and the following basic equation applies.

$$\mathrm{UST} = \frac{(\rho_s - \rho_f)\, g D^2}{18\eta} \tag{5.1}$$

where UST = the Stokes terminal velocity, ρ_s = the density of the particle, ρ_f= the density of the liquid, g = gravity, D = diameter of the sphere and η = viscosity of the fluid.

In true creep flow conditions the sedimentation of the particle will not alter its orientation within the liquid once a stable orientation has been achieved from rest. If the centre of rotation and the centre of mass are not in line with the force of gravity, an initial turning force will exist. The particle will then orientate until the centre of mass and of rotation are inline with the gravitational force.

Under creep flow conditions the drag on the particle is simply the sum of all elemental surface experiencing drag by the fluid. The downward force on the particle is determined by the difference in density between the particle and the fluid and the volume of the particle. Therefore, for any particle its volume determines the downward force together with the difference in density but its surface area determines the drag.

The sphere is a unique body since its shape occupies a maximum volume but a minimum surface area.

Imagine an irregular particle sedimenting at the same velocity as a 20 μm sphere of the same density. It will be ascribed the same size value as the sphere using equation (5.1).

As an irregular particle will have a greater surface area than that of an equivalent sphere, it follows that the drag will be greater. Therefore, if the irregular particle is sedimenting at the same velocity as the 20 μm sphere, it will certainly have a greater surface area, and its volume must be greater than that of the 20 μm sphere, to produce a greater downward force to compensate for the increased drag.

Stokes equivalent diameters tend to attribute a size to irregular particles smaller than they really are. This is particularly true for clay-like particles which are flat or curved discs of some 0.5–2 μm in diameter but may be only 10–20 Å thick.

The overall time taken to conduct a full measurement using sedimentation is directly related to the width of the distribution.

The time taken for increasingly small particles to reach a specific measurement point increases in proportion to D^2. If distributions are assumed to have particles from 100 down to 1 μm, and we further assume it takes 0.6 seconds to ensure the correct measurement of the 100 μm particle, then the total measurement will take two hours forty six minutes. The technique is generally slow, although current units do offer faster but less precise results.

5.2.3 Optical microscopy

The ability to view a system of particles under the microscope is a powerful and necessary facility for any particle sizing laboratory. Even before the use of the microscope for sizing purposes is considered, much useful information about the nature and structure of the particles to be studied can be extracted just by looking at them. The fact that the shape of the particle considerably influences the size that is ascribed to it by the various sizing methods used has already been noted. Confidence in the final state of the dispersion from lightly agglomerated forms can usually only be guided by observation of the target particles floating in a drop of fluid.

The optical microscope is used over a size range of about 1–150 μm. For particles which are smaller than 1 μm the electron microscope is used, and for simple inspection purposes a hand magnifier can be used for particles greater than 150 μm.

With particles lying on a suitable substrate, a two-dimensional representation of the particle is being observed, a feature which becomes more significant for flake-like particles of higher aspect ratio. It becomes increasingly difficult to determine the true edge of the particle at sizes below 5 μm due to diffraction effects. Reference to any text on sizing by optical microscopy will show a variety of ways of providing a single dimension of size for a two-dimensional object. The method chosen often depends on what comparison reticules are available or what feature of the particle size is most required.

After a size parameter method has been chosen, a distribution of the sizes present must then be formed. A number of size-classes are set, each having an upper and lower boundary. Each particle is observed in turn and its 'size' determined, adding one more to the size-class bracketing the observation. From many such observations a histogram of the number of particles which populate each size-class is formed.

A number of observations can be made at this point.

In order to be confident about the population of particles in any one size-class, to a standard deviation of better than 10%, 1000 particles in that one size-class alone would need to be counted. If a distribution is set over ten size-classes then we need to have seen and measured 10^4 particles. Clearly a need for automation is defined.

Many final results for particle size distribution are needed in terms of the volume or mass of particles in each size-class. 10^6 (1 μm) particles are required in order to obtain the same volume as a single 100 μm particle. Number and volume are related by D^3. Any errors in determining a number density distribution are greatly magnified when recalculating to determine a distribution by volume. For narrow fractions this is not a significant problem, but for samples which are polydisperse above about 5:1 in size range the errors rapidly increase.

The lesson here is surely that if a distribution is required by number, then each particle must be observed and counted. If, however, a distribution is required by volume or mass, a method must be chosen whose primary response is to volume or mass.

Automated image analysis software programs are available and in common use in some areas. The art in designing automated analysis is to answer three important questions.

- How does the hardware/software operation determine the true edge and hence the true boundary of the particle?
- What automated procedures are used to separate two objects which appear to be touching (are they really separated or not)?
- What rules are present to prevent the bias that exists with the larger particles being more likely to be excluded by touching the edge of the frame?

Particles which image with low phase contrast will naturally produce the most difficulties for automated systems.

5.2.4 Electrozone sensing

The origins of this technique stem from the need to count the proportions of red, white and platelet cells in whole blood.

The blood cells in an electrolyte are passed through a tiny hole (about 100 μm) in the glass wall of two separate chambers, each of which contains an electrode. If a small voltage is connected across these two electrodes a small current flows in the electrolyte, limited by the impedance of the hole $R = \rho\, L/a$ where L = length of hole and a = cross-sectional area.

As the blood cell or particle passes through the hole the impedance of the hole rises and an electrical pulse is generated. The amplitude of this pulse is proportional to the volume of the cell or particle. Therefore, both the volume and the number of particles can be determined.

The particles must be dispersed into the electrolyte at comparatively low concentrations to avoid two or more particles being present in the hole at the same time causing particle coincidence problems. The size range is limited to between 2 and 40% of the diameter of the hole, and to a rate of counting of about 5000 per second. The fact that the particle must be suspended in an electrolyte does raise some difficulties with respect to dispersion in particular, and to choosing a satisfactory electrolyte in general.

The technique is well capable of producing high resolution particle sizing within a dynamic range of about 20:1 for any specific hole for a wide range of materials. However, it is not satisfactory for materials of high density or high porosity. An excellent text on the foregoing methods is provided by Allen (1990).

5.2.5 *Laser light scattering methods*

The foregoing methods are still in use in many laboratories. However, the use of laser light scattering based systems has become more dominant lately because of their ease of use, and wide dynamic range of operation.

The particle size distribution of the cocoa powder used in chocolate, influences the customer's perception of its taste, together with the economics of its production.

A typical size distribution of a chocolate bar is shown in Figure 5.1 analysed by light scattering. The whole size range is smaller than is generally practical even for wet sieving. The small size range and low density difference when dispersed in trichloroethene makes sedimentation experiments protracted. The particles are too small and too polydisperse to make microscopy a viable route.

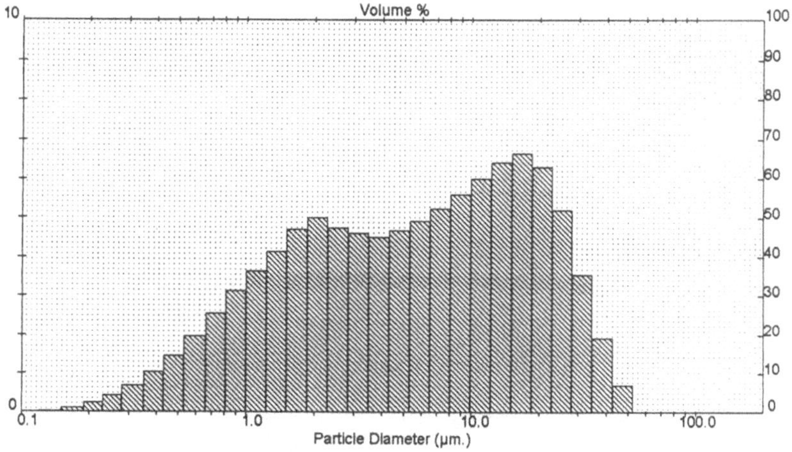

Figure 5.1 Typical particle size distribution for a chocolate bar.

For chocolate, the recognised dispersion medium is trichloroethene, which is not an electrolyte so that electronic sensing is not possible. Electrolyte dispersions have been used, together with small orifices, but the method is not one currently preferred.

It is now easier to see why a laser light scattering machine which can take measurements from 0.1 to 80 μm, with the particles suspended in any clear liquid, and measure in a rapid and reproducible manner, would gain acceptance. Equivalent situations have been seen with emulsions, starches and creams.

With this in mind, more attention will be paid to the light scattering methods.

5.3 Low angle laser light scattering (LALLS)

5.3.1 Introduction

The earliest instruments based on laser diffraction techniques (Swithenbank *et al.*, 1977) were first made available in the mid-1970s and represented yet another method of particle sizing. However, devotees of the technique were impressed by the ease of use, freedom from direct calibration and good reproducibility that the systems offered. As a result, the method steadily gained a foothold in the market place.

Early units from Malvern were employed in spray and droplet size analysis of fuels, pesticides, herbicides, aerosols and so on, where a more rapid acceptance was gained as no other easy and convenient method was available for this purpose. Later a full range of liquid suspensions units was added, together with dry powder feeders. The earliest units used a computer (PDP8) with 8K of core-store memory and a teletype as the human/machine interface.

Fraunhofer diffraction theory (Fraunhofer, 1821) which was comparatively simple to code into the computer was used exclusively, and thus the terminology for 'laser diffraction particle sizing' was born.

However, today use is made of the rigorous theory of Gustave Mie (1908) which allows for refraction, diffraction and absorption. Measurements are made over a series of low scattering angles from the laser beam and so the more complete title is needed.

The set-up of a typical LALLS instrument is shown in Figure 5.2. A monochromatic light source from either a gas laser (He–Ne) or a solid state diode laser is used. Visible light wavelengths have the advantage that the scattering from dirty optics or gas bubbles in liquid systems is readily seen by the user.

The laser beam is spatially filtered and expanded to form a collimated beam of some 9–18 mm diameter. A field of particles, either sprayed

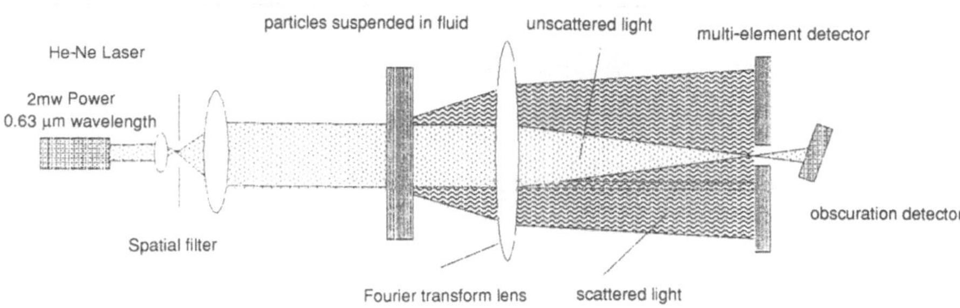

Figure 5.2 Components of laser diffraction particle sizes.

through the beam, or contained in a suitable cell assembly, is positioned in front of a positive lens. The unscattered light is brought to a waist of focus at the focal length where a multi-element silicon photodetector is positioned. In the centre of this detector a tiny hole is drilled allowing the unscattered light to pass through and to illuminate an obscuration detector. Light scattered by the particles is collected by the lens which acts as a Fourier transform lens.

An important property of this Fourier transform lens is that the scattering patterns produced by the particles are collected to form a stationary scattering pattern on the detector assembly, regardless, within limits, of the position of the particle or the fact that the particles are moving across the laser beam. The intensity of light falling in each detector element generates a small current which is amplified, digitised and read to computer memory. This data is then converted to particle size information which can be displayed on a visual display unit (VDU), stored onto magnetic media, or made available as hard copy via a printer.

5.3.2 Theory

5.3.2.1 Fraunhofer. The earliest instruments exclusively used the theory of Fraunhofer. This effectively assumes that all particles are spheres, fully opaque and that all sizes of particle scatter light with equal efficiency.

Figure 5.3. shows the basic equation and the resultant form of the scattering function predicted. This illustration has been for the scattering from a single particle. Clearly we need to store in the computer's memory the scattering pattern for the range of particle sizes covered by the focal length of each lens available. This is shown in graphical form in Figure 5.4. The numerical values of the predicted scattering pattern are stored as numbers in a 'T matrix' and labelled.

5.3.2.2 Anomalous diffraction scattering. The appearance of the first

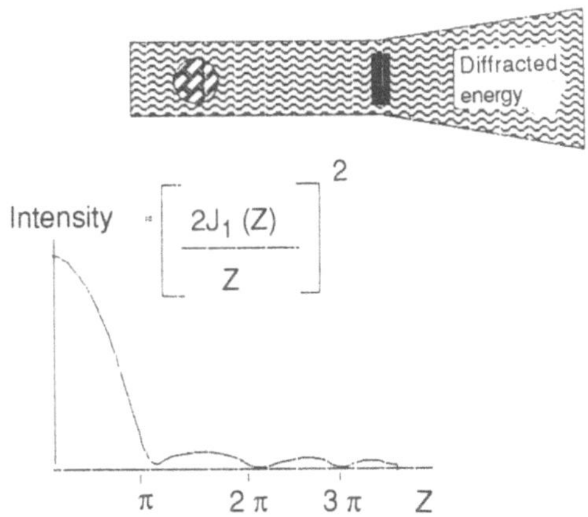

Figure 5.3 Fraunhofer theory assuming particles are fully opaque and all particles scatter with equal efficiency.

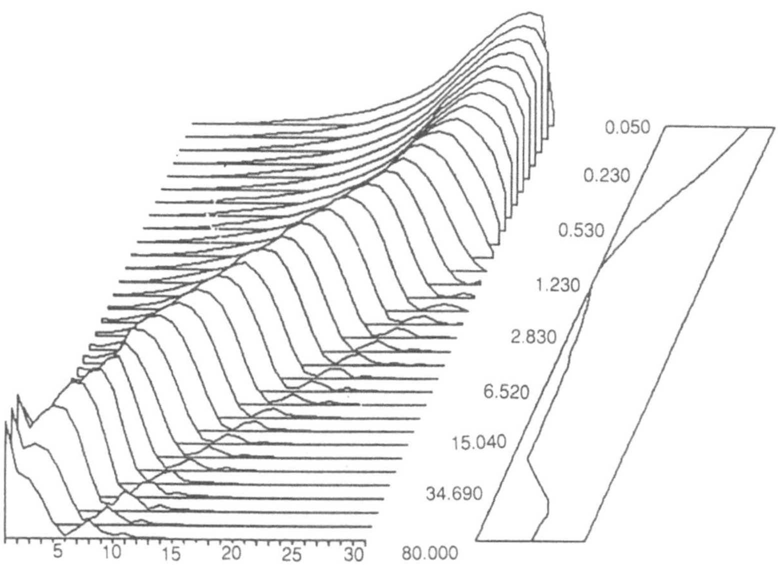

Figure 5.4 Fraunhofer diffraction scattering. Numerical values proportional to amplitude of each curve at each detector are stored in computer as a 'T'-matrix.

desk top computers, the Apple and the Commodore PET, not only made it possible to provide better graphical presentations of the data on VDU and printer, but enabled more complex optical models of scattering to be employed.

Many particles are not opaque and thus the extra light recorded by the detectors due to light passing through the particles as well as diffracting around it, produced less than satisfactory particle size distributions when analysed by the Fraunhofer model. The anomalous diffraction scattering equations assume spherical particles that are transparent but do correctly anticipate that small particles scatter less efficiently than larger ones.

This is still an approximation to the real world, but for the majority of industrial particles suspended in liquid the resulting particle distribution given is closer to reality. An example of the scattering predictions is shown in Figure 5.5.

5.3.2.3 Mie theory. Further advances in desk top computers both in their speed and memory capacity have finally made it possible to employ the rigorous solution to Maxwell's equation for the scattering of a homogeneous sphere set down by Gustave Mie in 1908. The equations are complex and numerically intense but are exact for spheres from molecules to footballs. The calculation time for Mie prediction for each size of particle increases with particle size, by approx. $D^{1.5}$.

In most cases the scattering prediction by Mie and Fraunhofer for particles above about 40 μm converge, therefore the faster Fraunhofer calculation is used above this value.

The Mie calculations for the scattering from a sphere for a three-dimensional object are positive. The geometric optics approximation of Fraunhofer who modelled the scattering of a thin disk gave rise to a misconception that the LALLS units responded to a projected area.

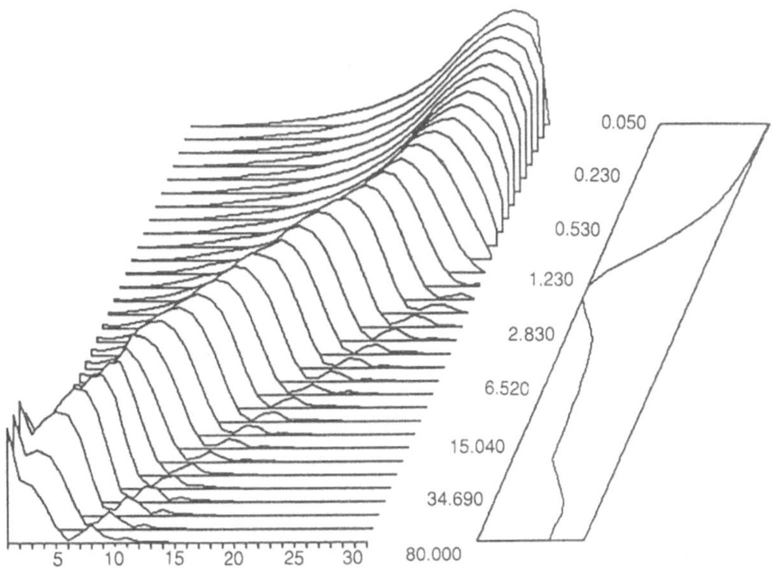

Figure 5.5 Anomalous diffraction scattering.

The two theories do converge for large opaque objects and thus would describe the two objects as having the same size. However, we could not ascribe the correct distribution of the volume of each size from a projected area model.

Measurements of the well characterised and certified distribution of solid glass spheres supplied in earlier times by the National Bureau of Standards (now called NIST) NBS1004 are reproduced in Figure 5.6 proving the volumetric response. Mie theory predicts the angular dependent scattering patterns, the effect of polarisation and the extinction, absorption and scattering efficiencies for each size of particle (Van de Hulst, 1957). It has been tempting for some equipment designers to use only part of the predictions in order to simplify matters, particularly for a limited range of small narrowly distributed spherical particles like latex, with refractive indices (RI) between 1.5 and 1.7.

The use of polarisation intensity difference ratios can be quite successful, but for non-spherical particles of general material these short cuts lead to ambiguous results, and are thus not realistic. The full theory needs to be applied.

5.3.3 Material refractive indexes

To utilise the Mie equations the RI of the material to be measured must be known. At first sight this may look difficult but patient search in many areas of the literature has enabled a library of data on the more commonly encountered materials to be established.

In cases where it is absolutely impossible to find or ascribe a reasonable value for RI then the non-optical property dependent model of Fraunhofer is still available, on the understanding that comparative results are being given rather than ones closer to reality. Even when the real part of the RI is

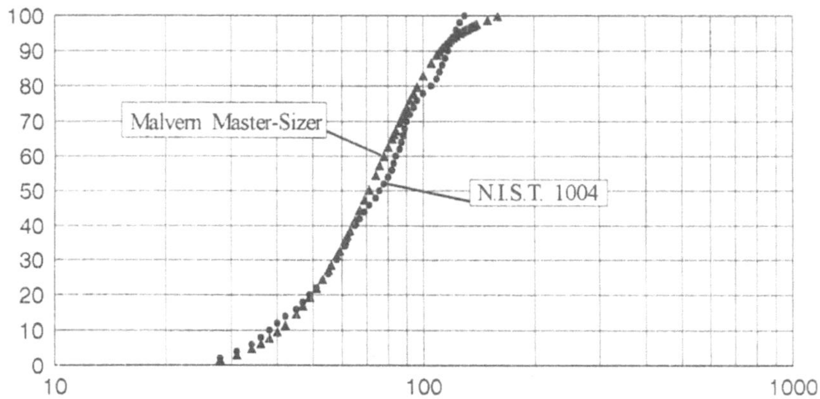

Figure 5.6 Standard reference material, NIST 1004.

known there may be a doubt over the complex part.

In recent work by Lips *et al.* (1992) known volumetric concentrations of sample are compared with the result of the volumetric concentration reported by the LALLS machine. The reported volumetric concentration uses the obscuration value as input to a Lambert–Beer calculation in which each size-class reported has assigned to it the predicted extinction coefficient which is then summed over the whole size range. Agreement better than 10% between the reported volumetric concentration and the known concentration of the sample gives positive confirmation that the correct optical model is being used.

Iteration of the complex part of the RI and to some degree, both the real and complex parts, can be used to define finally the values of RI for those systems which are known to be mixtures of materials and for which a single value of RI has to be given. Lips also investigated the effects of particle shape on the above approach, and found that it remained robust and reliable up to aspect ratios of 5:1.

5.3.4 Concentration range

The scattering by an ensemble of particles is assumed to obey the single scattering criteria. This is the state when scattering by a single particle passes to the detector without subsequent interference by other particles. A concentration is eventually reached whereby interference by multiple scattering processes takes place.

Particle sizing is more usually confined to the condition of single scattering but attention is now turning to handling the more complex problem set when higher concentrations do induce multiple scattering conditions. Figure 5.7 shows single scattering limits of operation for a LALLS machine.

The lowest concentration is set by signal over background conditions. The highest concentration is set just prior to the onset of multiple scattering.

Figure 5.7 shows us that the volumetric concentration for single scattering reduces as the particle size reduces. This is loosely because the number density distribution of particles per cubic centimetre of fluid governs the onset of multiple scattering. Clearly there is a greater number of small particles for a given volume concentration of sample than larger ones.

5.3.5 Data analysis

The introduction showed the need to compare the scattering from a system of particles with a theoretical model of the way such particles scatter light. The instrument measures the resulting light energy distribution as a result

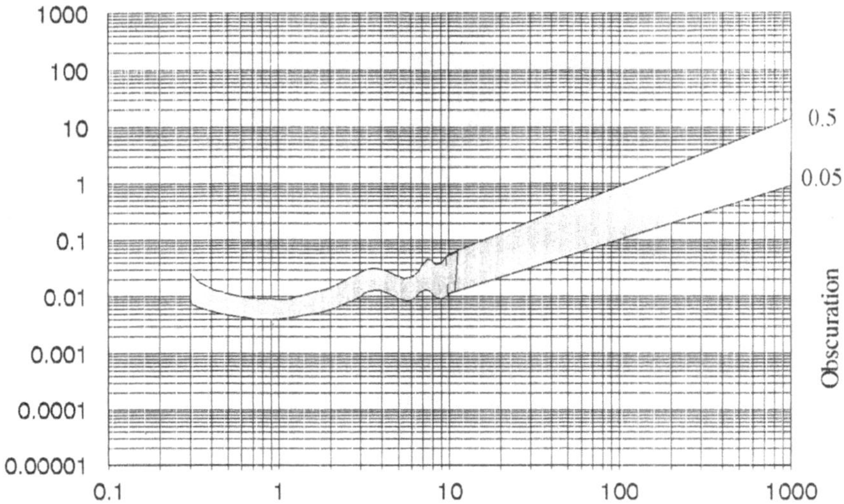

Figure 5.7 Relative concentration. Volume concentration range for Mastersizer presentation 0807.

of this scattering, but what is required is the volumetric distribution of particles that could have created it.

Bohren and Huffman (1983) expressed this rather neatly. The sort of tracks a dragon would leave can be readily imagined by studying its feet. However, here the tracks of the dragon are available and the need is to deduce what sort of creature created them.

Mathematicians describe the inversion of these particle scattering functions as ill-posed and ill-conditioned. This is their way of saying that there are going to be limitations on what can be expected.

When taking the signal from an ensemble of particles and breaking it down to the component distribution there will always be limits to the resolution that can be expected. Resolution is not related to the number of data channels displayed, as some advertising implies. It is, however, related to the system's ability to discern the separation of two narrow fractions, Figure 5.8.

The greatest advance of the LALLS machines has been their ability to cover a wide range of sizes (typically 800:1 plus). However, when inverting integrated data there is an unspoken law which says 'you can either have dynamic range or you can have resolution, but not at one and the same time'.

The final solution adopted by each software engineer faced with this problem is, in the final analysis, one guided by text book approaches, but tempered by field experience. If there is *a priori* knowledge that narrow

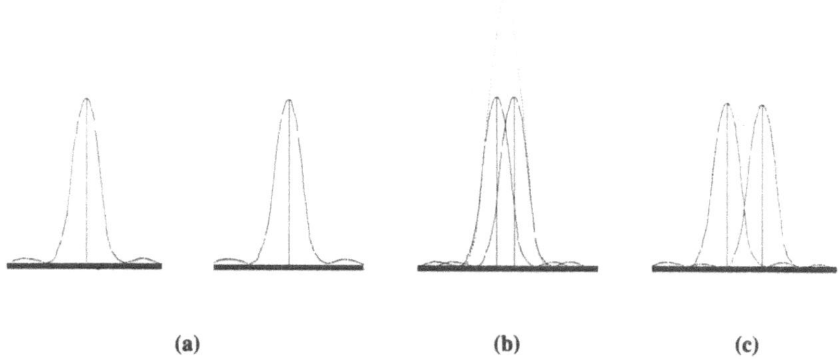

Figure 5.8 Separation of two narrow fractions. (a) two well-separated Gaussians, (b) two overlapping Gaussians which are not resolved, and (c) two Gaussians which are just resolved.

distributions of particles are involved, only then does it become justified to seek a higher resolution solution over a limited dynamic range, Figure 5.9. Mastersizer software allows this.

5.3.6 Criteria for a minimum sample

'The volume of sample required to be fully representative of any material increases with increasing range of particle sizes present'. This general truth can be illustrated by examining the limits.

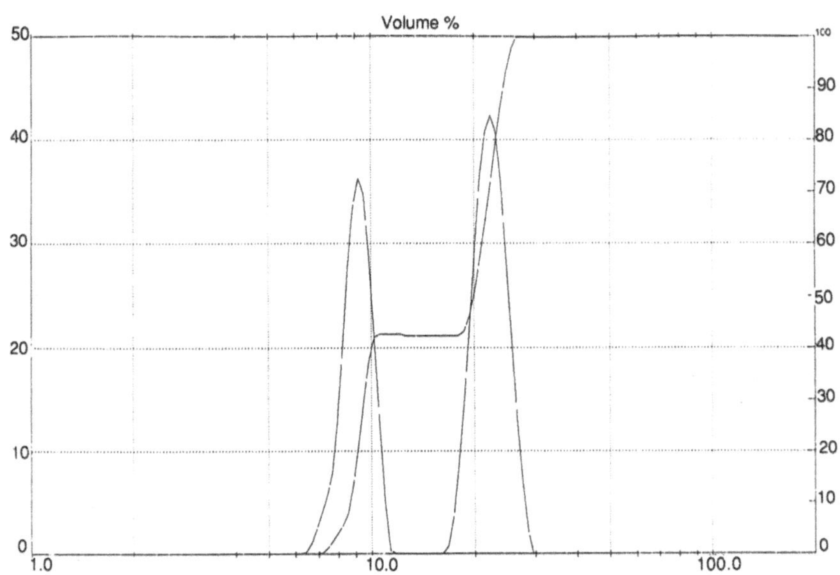

Figure 5.9 High resolution result. Mixture of 21.1 and 9.33 micron latex.

A truly mono-sized particle system only requires one particle to be sampled to know the characteristics of all the other particles present. Thus the volume of one particle is the minimum volume of sample required.

Conversely, a geologist assessing the particle size of a glacial moraine would be faced with sub-micron colloidal suspensions on the one hand and boulders of several cubic metres on the other, with all sizes in between.

To replicate adequately a sample of such a glacial moraine, a sample of many hundreds of tons would be needed to have any chance of assaying the size distribution which contained the largest elements to any degree of statistical accuracy.

Clearly everyday samples fall between these two extremes and so the following simple approximation is proposed as a guideline.

5.3.7 *Minimum sample volume assessment*

A typical size distribution is shown in Figure 5.10 where sizes close to 200 μm and down to below 1 μm are present. Consider the largest particle (*LP*) to be 200 μm in this case.

In repeated sub-samples of the material under study, further assume that this largest fraction must be sampled to a coefficient of variance of 1% or better. This then predicts that a minimum of 1000 particles of 200 μm would be needed statistically to achieve this coefficient of variance.

Let the volume of these largest particles be 1% of the total distribution then the minimum sample is

$$\frac{\pi\,(LP\,10^{-4})^3}{6} \times 1000 \times 100\ \% \tag{5.2}$$

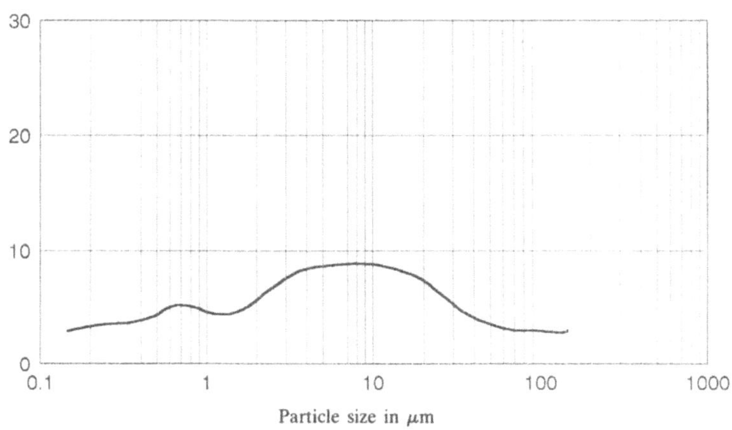

Figure 5.10 Polydisperse size distribution.

Multiplying by density gives the minimum weight.

As an example assume the distribution of Figure 5.10 is made of crushed quartz, density 2.6 g cm^{-3}, then the minimum volume for this example is 0.42 cm^3 with a weight of approximately 1.1g.

5.3.8 *Sample dilution*

If such a sample is to be used in a low-angle light scattering unit we must consider the highest concentration permissible before multiple scattering effects impose a reduction in accuracy.

Figure 5.7 shows a general curve of the volume concentration range over which acceptable results can be expected. Attention must be paid to the expected presence of particles in the region of 1 μm where a middle range relative volume concentration of 10^{-2}% should be aimed for. 0.42 cm^3 of powder must therefore be dispersed in a minimum of 42 cm^3 of dispersant. This simple illustration indicates that any sample smaller than the minimum indicated is likely to result in more uncertainty in the sampling of the larger size fractions, and that a minimum total volume of dispersed mixture is also needed.

5.3.9 *Results*

From the data analysis a preselected number of histogram values representing the percentage volume in each of the size classes selected, is given. These values are converted into a cumulative undersize curve. From this primary data set, a curve fit is performed, and thus a law representing a continuous curve through the discrete number of data points is established.

It is now possible to calculate or select a vast array of percentiles and

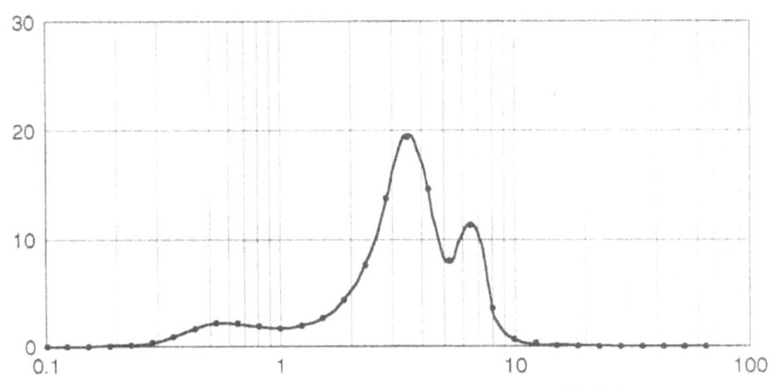

Figure 5.11 Long life milk.

moments of the distribution from which end users select a limited set, dependent upon their requirements.

As has been explained, the primary response of the instrument is to the volume of material in each size-class. It is an easy matter for the computer to transform this volume distribution to one of number. However, just as in the case of microscopy, the errors transform in a cubic manner, rendering the number distribution prone to distortion.

Figure 5.11 shows the particle size distribution of a dispersed long shelf life milk. The molecular weight of the protein in the milk is such as not to contribute to the scattering. Therefore the distribution of fat globules is being sized.

5.4 Sub-micron particle sizing

For particle sizing of objects above 0.1 μm the angular dependence of scattering to characterise the particle can be used. Figure 5.12 shows that the angular dependence for particles less than 0.1 μm becomes more usable for sizing purposes.

In the late 1960s digital correlators became available stimulated by research into signal recovery problems in radar and other noise limited detection situations. Low-level light scattering experiments are often detected using photomultiplier tubes. Light levels of detection are such

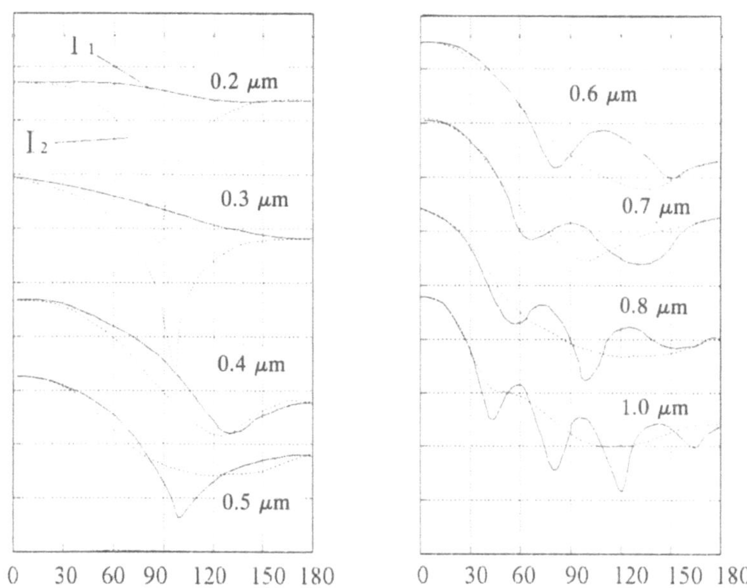

Figure 5.12 Scattering of non-absorbing spheres.

that the absorption of light energy by the photocathode is only sufficient to liberate single electrons.

Thus we are detecting single photon events. The statistics of photodetections are such as to be much better suited to digital autocorrelation for signal processing than to other spectral analysis techniques.

With the combination of laser and digital correlators the stage was set for the introduction of photon correlation spectroscopy (PCS) for particle sizing. It is worth noting that PCS has also attracted many other descriptive titles (Pusey *et al.*, 1972) all of which attempt to capture the essence of the technique in a single title.

5.4.1 Principle of operation

A typical optical set-up is shown in Figure 5.13. A glass or plastic cell containing a dispersion of molecules or particles is placed in the laser beam. Each individual particle scatters light. The radiation pattern in the case of particles less than gk $\lambda/20$ is isotropic. The detector is positioned at some angle to the incident laser beam (usually 90° for convenient operation).

Light scattered by each particle has the same wavelength as the incident beam and acts as a coherent source. Therefore, the detector records an intensity which is the phase addition of light scattered by each particle. A screen placed at a low angle would show a 'speckle' pattern. The bright spots or speckles are where the phase addition of the scattered light is

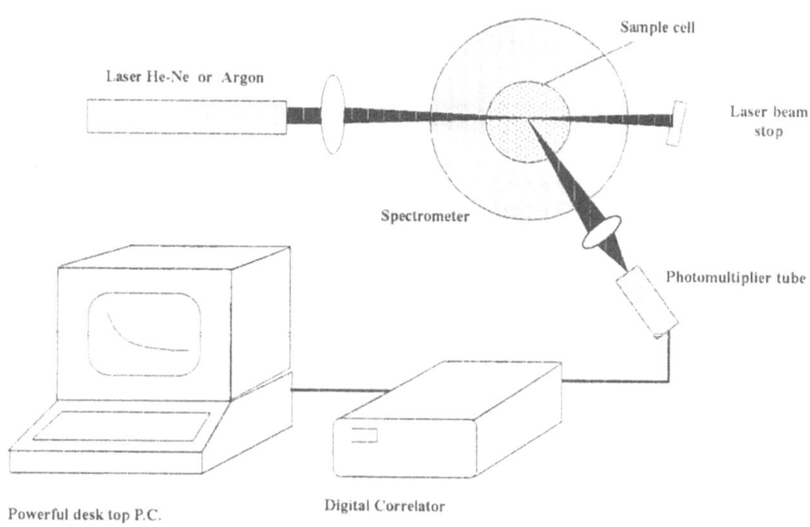

Figure 5.13 Typical optical arrangement for PCS.

constructive and adds. The dark spaces between the speckles are where the phase additions cancel each other out.

Particles suspended in a fluid are bombarded with the electrons of the solvent and are made to move. This 'Brownian Motion' was first reported by the Rev. Brown, who noticed that tiny pollen grains suspended in water were in 'perpetual motion'. It was not until the work of Stokes and Einstein that a satisfactory and rigorous explanation was given for this motion.

Because the particles are undergoing Brownian motion, the speckle patterns created by the scattered light are also moving. Therefore, if a detector aperture equal to the size of one of these speckles (coherent detection) is chosen, then it ranges from being either fully illuminated or dark in an apparent random manner. The signal output from the photomultiplier is therefore fluctuating in intensity at a rate characteristic of the rate of movement of the particles.

The time constant of the exponential autocorrelation function which is observed as a result of this temporal fluctuation, relates directly by a simple equation to the rate of translational diffusion of the particle. Using the Stokes–Einstein relationship, the hydrodynamic radius of a spherical particle can be calculated, and a particle size determined.

5.4.2 The analysis problem

If all the particles are the same size, they will diffuse at the same speed on average, and an exponential autocorrelation function with a single time constant will result. However, most practical particle sizing situations require study of a polydisperse range of particle sizes. Each size will diffuse at a different rate so that the resulting autocorrelation function will be a series of different time constant exponentials, all summed together to form a non-exponential exponential.

Again we face an inversion problem to deconvolve the number of exponential time constants and to assign the relative magnitudes of each one. This is an even more challenging problem than the one faced by the LALLS machine designer.

Each data point on the autocorrelation function has some uncertainty in its value. This results from photon statistics and an unquantifiable amount of correlated and un-correlated noise. The photon statics can be reduced by observing a greater number of photons, i.e. by running the data collection of the experiment for longer.

However, the longer the experiment takes the more it becomes vulnerable to tiny changes in laser intensity, photomultiplier sensitivity, the migration of aggregates into the path of the laser beam, and many other tiny irritants.

The net result is to limit the resolution of particle size distributions. Further, for particles smaller than about 0.1 μm their intensity of scattering reduces as R^6. This means that individual particles of 0.2 μm scatter one million times as strongly as particles of 0.02 μm. This element in the physics of light scattering is a further restriction on the range of sizes that can be measured at any one time.

5.4.3 *Results*

Despite the restrictions the technique is highly valuable for colloids, micelles, lipids, proteins, microemulsions and sols, and so on.

Various sizes of latex particles have been certified in terms of their size and thus it is common to see the results for these materials in the literature, Figure 5.14.

In a series of experiments examining the micelle structure of milk (Figure 5.15) data on the micelle size and charge (see section 5.5.3) have been obtained both before and after the addition of rennet to assist in the understanding of cheese production.

The primary datum from PCS measurements is a size which has been deduced based upon the intensity of the scattered light. It is now known that small particles scatter more weakly than large ones in general, and thus the observations are intensity biased.

In order to relate more closely to the volume or mass of particles of each size, the intensity bias must be un-weighted. Here again Mie theory can be useful as it predicts the intensity of scattering per particle at all angles. Indeed, for particles greater than about 0.2 μm the angular intensity distribution becomes pronounced and contain peaks and troughs as found in the LALLS studies. Therefore, for particles above about 0.2 μm it is desirable to make a series of measurements at several angles to add further

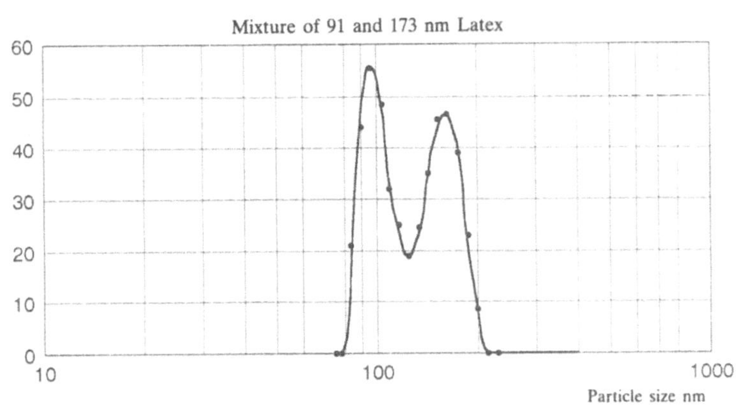

Figure 5.14 Measurement of latex standards by PCS.

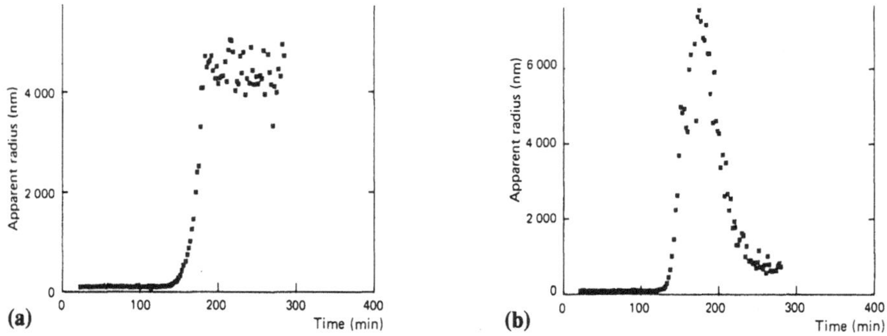

Figure 5.15 Gelation of renneted milk: (a) increase in apparent radius of particles as function of time after adding 33 μl of rennet (T = 30°C, starter = 6g per 10l milk), (b) change in apparent radius with time in experiment dominated by effect of adding rennet (T = 33°C, rennet = 66 μl, starter = 6g per 10l milk).

parameters to the intensity fluctuation data, in order further to refine the ability of the PCS technique to yield particle size distributions (Figure 5.16).

Many particle systems are at much higher concentrations than that suitable for single scattering. It is also true that some emulsions are unstable upon dilution. For these situations a fibre-optic arrangement (Figure 5.17) has been produced to enable direct measurements at concentrations between 0.2 and 50 % to be made under conditions representing real particle concentrations.

The concentration may well produce hindered diffusion, and thus the interpretation of diffusion to a true particle size might be difficult. However, for process or production control circumstances a relative measurement which changes in sympathy with product changes, is highly valuable. Dilution to a single scattering state for stable systems can also be carried out to obtain the true particle size.

5.4.4 Required practice for PCS measurements

The controlled dispersion of a system of particles into scrupulously clean solvent is a vital ingredient for successful measurements in PCS. Without good housekeeping and meticulous attention to ensuring cells, pipettes, stoppers and any other particle handling systems are clean, success will be elusive. The use of in-line filters of about 0.25 μm to clean the solvent and rinse the cells and stoppers, is highly recommended. With some materials dispersion can only be guaranteed in the correct buffer medium which itself will need filtering.

The concentration of particles to be measured must again be confined to that which provides single scattering. If the dispersed particles are illuminated with the laser beam and produce a crisp single line of scattering then

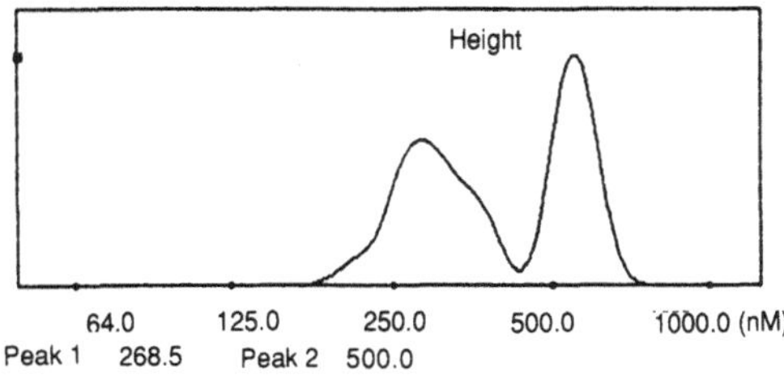

Size (nM)	Number (%)	Volume (%)	Intensity 45	Intensity 70	Intensity 90
50.0	0.00	0.00	0.0	0.0	0.0
64.0	0.00	0.00	0.0	0.0	0.0
80.0	0.00	0.00	0.0	0.0	0.0
100.0	0.00	0.00	0.0	0.0	0.0
125.0	0.00	0.00	0.0	0.0	0.0
160.0	0.00	0.00	0.0	0.0	0.0
200.0	20.46	5.68	2.0	7.1	8.3
250.0	52.84	28.68	17.5	50.9	48.2
320.0	15.68	17.85	18.5	35.9	19.8
400.0	0.00	0.00	0.0	0.0	0.0
500.0	11.00	47.77	62.0	6.1	2.37
620.0	0.00	0.00	0.0	0.0	0.0
800.0	0.00	0.00	0.0	0.0	0.0
1000.0	0.00	0.00	0.0	0.0	0.0
1250.0	0.00	0.00	0.0	0.0	0.0

Figure 5.16 Multi-angle measurement PCS data. Mixture of 250 nM + 520 nM latex measured at 45°, 70° and 90°.

single scattering is being achieved. If, however, the scattered beam appears surrounded by a halo or is a steadily diverging beam then multiple scattering is evident and dilution is needed.

The theory assumes the particles are in infinite dilution, i.e. no particle–particle interaction is taking place. Even with single scattering seemingly assured, it is good practice to make at least one or possibly two dilutions of the material to test if the diffusion coefficient changes with

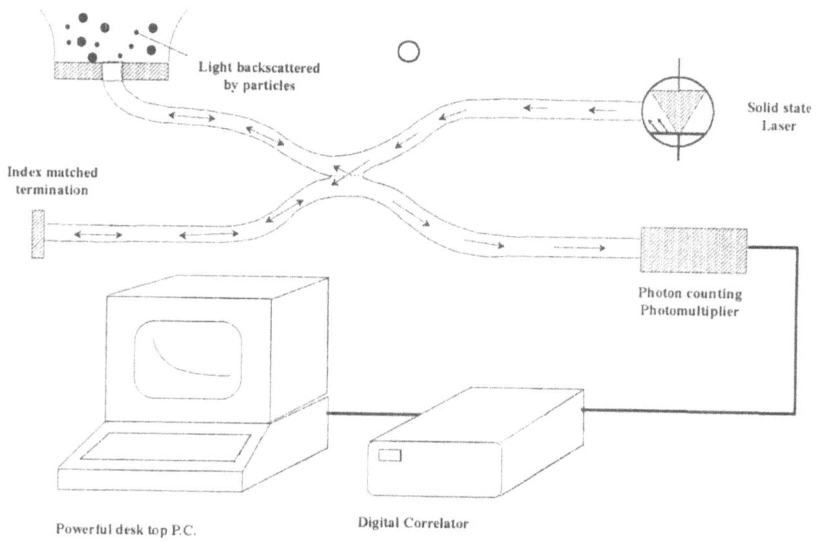

Figure 5.17 PCS fibre-optic layout (Malvern HI-C).

concentration. The amount of laser intensity needed will, as we have seen, depend upon the size of the particles and their scattering power and secondly the concentration which permits free diffusion. For most particles above 50 nm a 10 mW He Ne laser is all that is required. However, for smaller particles a more intense argon laser may be required. The environment of the equipment is also important. Mechanical vibration caused by other laboratory machines will require isolation as the technique is also a sensitive seismometer.

5.5 Measurement of mobility and zeta potential

5.5.1 Introduction

In the preceding discussion we have seen the importance of dispersion in obtaining good reliable particle size distributions.

When in flood the river Severn carries a burden of silt to the sea. This suspension remains stable for the journey. Most of the particles in the fresh water are negatively charged and repel each other. However, in the tidal estuary fresh water mixes with salt water. The positive ions attach to the negative charge sites and deplete the charge until particles can approach each other to form flocs, which increase in size until they precipitate out, and form the mud banks so characteristic of that river mouth.

This illustration neatly demonstrates that an additional substance, in this case salt, can destabilise the suspension. Conversely, the addition of a small amount of surfactant can be made to stabilise an aqueous based system of latex particles. An indicator of surface potential or charge is, therefore, valuable.

5.5.2 Measurement method

A typical set-up for the measurement of mobility by the laser doppler method is shown in Figure 5.18.

- A laser beam is split into two equal paths and made to cross through one another at the point of measurement. At the intersection of the two laser beams a set of Youngs interference fringes would be formed onto a screen. The spacing of the fringes is set by the equation $s = \lambda/\sin(\theta/2)$ where λ = the wavelength of light and θ = the angle between the two beams.
- The measurement cell has electrodes at either end of a capillary tube. A potential, typically 40–100 V is applied.
- If the cells are negatively charged they will be attracted to the positive electrode. If the cells carry no charge they will not experience the electric field at all.

Electrophoretic mobility is defined in $\mu m\ s^{-1}\ V^{-1}\ cm^{-1}$. It is easy to measure the field ($V\ cm^{-1}$). All that remains is to measure the velocity of the particles as they migrate towards the electrode of opposite charge.

Particles moving through the intersection of the two laser beams scatter

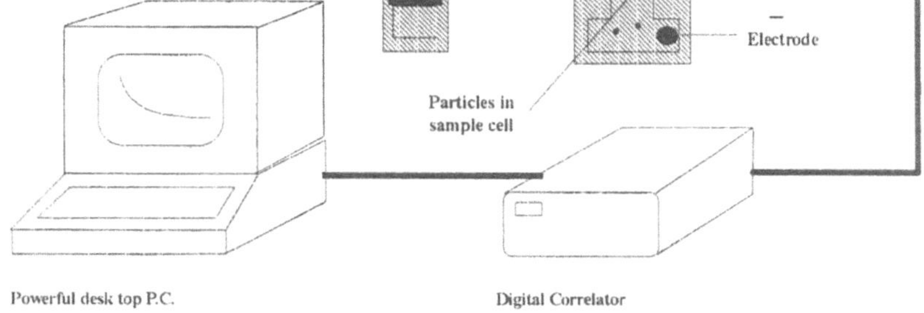

Figure 5.18 Schematic layout of zeta 4.

light which is detected by the photomultiplier tube. The light scattered by the particles from each of the laser beams is Doppler shifted as a result of the particle's movement. The cathode of the photomultiplier tube responds to the Doppler difference frequency, f, of the Doppler shifts from each beam. This is numerically identical to the simplified model of 'fringes'

where $$f = \frac{U}{s} = (2U \sin \theta)/\lambda \tag{5.3}$$

where U = velocity.

In addition to the electrophoretic mobility there is also Brownian motion.

As the single photon limit with coherent detection again applies here, a digital correlator is the signal processor of choice. The autocorrelation function is now a decaying cosine wave where its period represents the time taken by a particle to move one fringe pitch.

The autocorrelation functions and the frequency power spectrum are Fourier transform pairs. Therefore, after removal of the beam profile, a software executed fast Fourier transform (FFT) can be applied to yield the spectrum of Doppler difference frequencies. Scaling the frequency by s, the fringe spacing, gives a spectrum of mobilities from which a mean mobility can be calculated.

The width of the mobility spectrum is broadened by the effect of Brownian motion. With the voltage supply to the cell electrodes switched off only the Brownian motion is present which can be measured separately.

A second measurement is made with the cell electrodes switch on when a combined mobility and Brownian motion spectrum is observed. Spectral subtraction removes the effects of the Brownian motion, leaving a high resolution mobility distribution (see Figure 5.19).

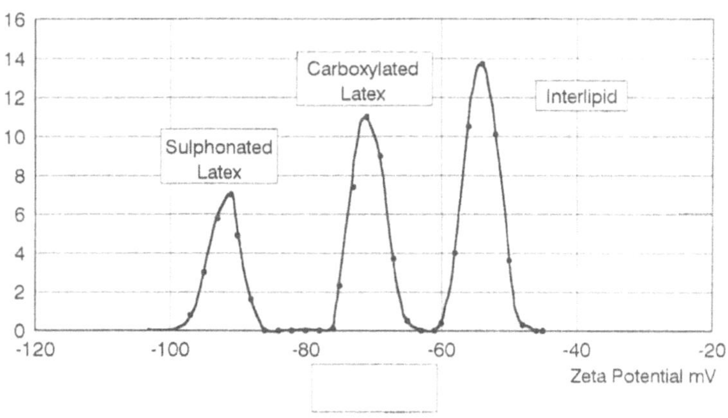

Figure 5.19 Separation of tri-mixture.

Having established an electrophoretic mobility, one of the appropriate models which link mobility to zeta potential can be chosen and the zeta potential calculated and reported.

5.5.3 *Applications*

Both the Brownian motion which yields particle size by the PCS technique, and the mobility of zeta potential, can be measured within the Zetasizer. A study of hematite was undertaken where the zeta potential was measured against pH of the suspending media (Figure 5.20). When the zeta potential is zero an isoelectric potential is reached where the tendency of the particles to aggregate is at a maximum.

An amount of dry hematite which would ensure dispersion at the correct concentrations for particle size measurement was made up in dispersion at various pH values. The dispersion procedure, followed by a standing time for thermal and mechanical equilibrium to take place, over a total fixed time was applied to each sample (2 minutes). After this period a measurement of particle size was made and the results plotted against pH.

At values of pH close to the isoelectric potential, aggregation raised the mean particle size to a maximum in the 2 minute procedure allowed. Stable dispersions could only be guaranteed at either very high or very low pH values.

The zeta potential of the micelles of milk plotted against the time after rennet was added to the milk is shown in Figure 5.21. Not only does the zeta potential of the micelles fall but their size also changes after rennet is added.

This work was carried out as part of a study into cheese manufacture by Dr. Horne of the Hannah Dairy Research Institute.

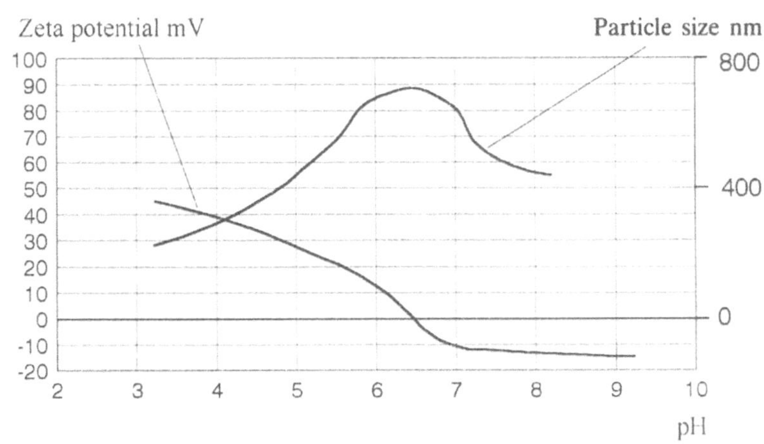

Figure 5.20 Hematite charge and size versus pH.

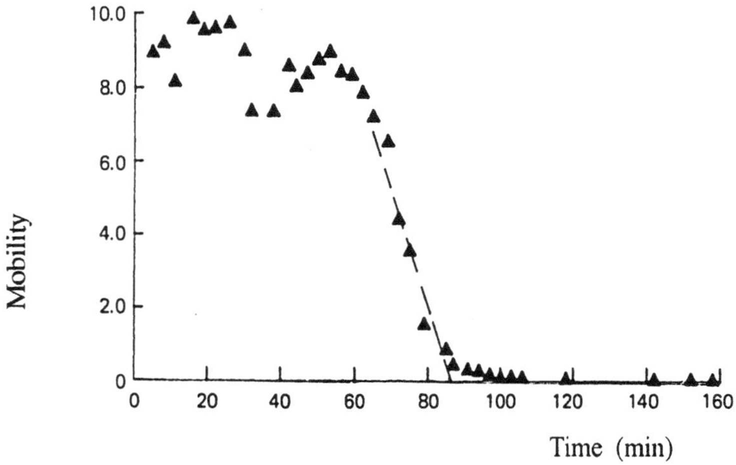

Figure 5.21 Rennet curd formation. Variation of charge on skim milk after addition of rennet.

References

Allen, T. (1990) *Particle Size Measurement*. 4th edn, Chapman and Hall, London.

Bohren, C.F. and Huffman, D.R. (1983) *Absorption and Scattering of Light by Small Particles*. Wiley—Interscience, New York.

Lips, A., Hart, P.M. and Evans, I.D. (1992) Use of low angle laser light scattering in characterisation of non-spherical particles and swelling of microgel particles. *European Symp. Particle Characterisation*. Nurenberg, March 24–26, pp. 443–457.

Mie, G. (1908) *Ann. Phys.*, **25** (3), 377.

Pusey, P.N., Schaefer, D.W., Koppel, D.E., Camerini-Otero, R.D and Franklin, R.M. (1972) *J.Phys.*, **33**, C1.

Swithenbank. J., Beer, J.m., Taylor, D.S., Abbot, D. and McCreath, G.C. (1977) *Prog. Astronaut. Aeronaut.* **53**, 421.

Van de Hulst, H.C. (1957) *Light Scattering By Small Particles*. Wiley, New York.

Von Fraunhofer, J. (1821–1822) *Dekscri Acad. Wiss., Munich*, **8**, 1.

6 Ultrasonic methods

C. JAVANAUD and M.M. ROBINS

6.1 Introduction

Ultrasound is defined as a longitudinal pressure wave with frequency above the range of human hearing (16 kHz). Uses of ultrasound include underwater range-finding (Anon, 1988) and medical applications (Hill and ter Haar, 1982). There are, however, other less well-known applications. Between 20 kHz and approx. 100 kHz, high power devices are used to initiate or speed up chemical reactions. This branch of ultrasonics is known as sonochemistry (Mason and Lorimer, 1989), and to date has found little application in food materials. However, high frequency ultrasonic methods (1–100 MHz) which usually operate at low power are used to examine food and related systems. The emphasis in these applications is on the use of low power to probe the material with minimal disruption. Although in principle the properties of materials are affected by their measurement, in practice, high frequency ultrasonic waves leave materials essentially unchanged. The great advantage of ultrasonics compared with other non-intrusive methods such as light scattering, is that the majority of food materials that are optically opaque do transmit ultrasound. The exceptions are systems containing dispersed air, since high frequency sound is strongly attenuated by gases. There remains a wide range of liquid based foods, and even certain solids, which are amenable to analysis using ultrasonic methods.

The propagation of ultrasound through a system depends upon its response to rapid pressure fluctuations and, in principle, all the mechanical and thermal properties have an effect. Foods are rarely homogeneous, and the transmission of ultrasound through a multi-phase material is influenced not only by the properties of the various phases in isolation, but also by the physical structure (concentration, size, distribution of subregions or particles) and the mismatch in ultrasonic properties between phases. The full picture is very complex, but the corollary is that the ultrasonic response is sensitive to many of the key properties of foods. There is thus considerable potential for using ultrasonic techniques to monitor the property of interest, provided that the samples are otherwise well characterised.

In this chapter, we present the fundamentals of diagnostic ultrasonics, in particular where they are relevant to food materials. Initially, in section 6.2, we summarise the principles of ultrasonic wave propagation, with special reference to multi-phase materials. Section 6.3 discusses how measurements may be made in practice, with indications of the precision and versatility of the various techniques reported in the literature. The application of ultrasonics to foods and related materials is reviewed in section 6.4, and we attempt to foresee future developments in Section 6.5.

6.2 Principles of ultrasonic wave propagation

6.2.1 General principles

The principles of ultrasonic wave propagation through single-phase materials can be presented in a simple way, provided that certain assumptions are made. When the propagation of ultrasound through materials containing two or more phases is being considered, even the simplest treatment is already quite complex. This is the case, whatever the nature of the multi-phase material, (whether, for example, it is a dispersion or a structurally much more complicated material). However, for certain combinations of material properties a simplification of the treatment may be possible, if the structures occupy particular size ranges, relative to the wavelength of the ultrasound.

It is important to be aware from the start that knowledge of any basic assumptions and/or simplifications which need to be made in the theoretical treatment are not just of academic interest. Failure to have sufficient knowledge of what is going on, and which equations are the most appropriate ones to use, will, at best, lead to reduced accuracy in the determination of measurable quantities. At worst, it may make a whole project worthless. Unfortunately, the scientific literature in this field contains many examples of experimentalists who seem to have done just that. One aim of this chapter is to help workers to avoid this in future.

6.2.2 Pure materials

The features of ultrasonic wave propagation of which most use has been made experimentally are the velocity and the attenuation of the wave. These quantities are best discussed first in connection with a single-phase material. It is necessary to assume that the ultrasound is produced at only a single frequency, f, and that it is in the form of a plane wave. (The former constraint means that only continuous waves are considered, while the latter means that the wave extends to infinity in planes normal to the direction of propagation.) The effect of relaxing these restrictions will be discussed in the next section. It is further essential to restrict the power carried by the wave, so that the maximum change in the density of the medium at any point in space, is much smaller than the average density at that point. Although this condition too, may be relaxed, the consequences of doing so are not considered here, as they are not relevant to any of the applications to be discussed in section 6.4.

Subject to the above restrictions, it is easily shown (Coulson, 1955) that a longitudinal ultrasonic wave propagates through the medium, in the x-direction with an amplitude A given by

$$A = A_0\exp(-\alpha x).\exp[i(kx-\omega t)] \tag{6.1}$$

where A_0 is the amplitude at $x = 0$, α is the attenuation in either Nepers m^{-1} (= m^{-1}) or db m^{-1}, $k = 2\pi/\lambda$ is the wavenumber, λ = wavelength, and ω = angular frequency = $2\pi f$.

The phase velocity at which the sound travels is $v = \omega/k$. The phase velocity is related to the modulus M of the material through which it travels by

$$v = \left(\frac{M}{\rho}\right)^{1/2} \tag{6.2}$$

where ρ is the density of the medium.

For fluids, the modulus M is the bulk modulus B of the material. For *isotropic* solids, M, is given by

$$M = B + \frac{4}{3}G \tag{6.3}$$

where G = shear modulus.

In solids, in addition to longitudinal wave propagation there is always an accompanying transverse shear wave, with the velocity given by equation (6.2), where $M = 4G/3$. For anisotropic solids, (including most metals and salts), it is not generally possible to separate the wave propagation into a purely longitudinal and a purely transverse component. In these cases, one cannot use equation (6.3) for the velocity of 'longitudinal' sound waves, as they do not exist.

For all the work of interest here, the moduli involved are the adiabatic ones, although, at very high frequencies, it is necessary to use isothermal moduli instead. The speeds of longitudinal sound waves through liquids and solids, at room temperature, are typically in the ranges 1000–2000 m s^{-1} and 2000–7000 m s^{-1}, respectively.

Many foods, or food related materials, do not behave ultrasonically as either pure liquids or pure solids; instead, they display viscoelastic properties. Such properties can be accounted for if *frequency dependent* bulk and shear moduli are attributed to the material. These moduli are a mathematical expression of the fact that such materials flow when subjected to slow strain, although under a rapid blow they behave like solids, some even fracturing.

Whereas knowledge of the wave speed provides information about the modulus of the material, the attenuation coefficient, α, depends on other material properties. Even in a pure material, there are many possible causes of attenuation, most of them being due to absorption of some kind. Heat conduction causes absorption to some extent in all media. An important cause of absorption in liquids is their viscosity. At comparatively low frequencies, which usually include the ultrasonic range, these two mechanisms lead to an attenuation coefficient which is proportional to the square of the frequency.

Liquids also absorb sound energy to excite their molecular energy levels and to undergo structural changes. These so-called 'relaxation' processes all follow a similar frequency dependence, if they can be normalised with respect to a single relaxation frequency. Otherwise it is necessary to superpose the contributions from many processes, each with its own relaxation frequency. In each relaxation mechanism, there is a simple equation governing the rate of energy loss from the sound wave as a function of the timescale required for the particular phenomenon to take place. Furthermore, the increase in attenuation near relaxation frequencies is always accompanied by a corresponding change in the speed of sound. The latter phenomenon is known as dispersion.

Apart from heat conduction, solids generally possess different loss mechanisms from liquids. Of particular interest here is the loss due to scattering from the individual grains in metals. Scattering differs from absorption in that the sound energy is redistributed in all directions by the scatterers. Thus, it is only a loss mechanism as far as the propagation of the incident wave is concerned. Consequently, if scattering alone is present, the total sound energy emerging from the medium will be equal to that which goes in, although it will no longer all be in the form of a plane wave. Scattering, like absorption, is a highly frequency dependent process. As discussed below, scattering is an important cause of attenuation when ultrasound travels through multi-phase media. The other loss mechanisms from solids will not be discussed here, as at the frequencies of interest, i.e. 0.1–1000 MHz, the attenuation coefficients of most solids are orders of magnitude less than those of most liquids.

6.2.3 Multi-phase media

The subject matter of the propagation of sound through multi-phase media is truly enormous. In this chapter, it is possible to provide only a glimpse into this subject. A fairly detailed mathematical approach, which also covers electromagnetic wave propagation, has been provided by Ishimaru (1978).

It is important to be clear about what happens as an ultrasonic wave propagates through a two- (or multi-) phase medium. To this end, it is easiest to start by considering the case where particles of one material are distributed throughout a second material, with the second medium forming a continuous phase. (The effect of modifying this geometry will be touched on later).

The plane incident wave, which is *coherent*, travels through the medium, being attenuated according to equation (6.1) where α, the attenuation coefficient is the sum of coefficients due to absorption and to scattering. The absorption coefficient required here is a coefficient allowing for both absorption in the background material and absorption by the particles themselves. There are essentially two contributions to particulate absorption. One is bulk absorption and the other is a surface phenomenon, closely resembling scattering. In the case where the scattering coefficient is much less than the absorption coefficient, the effects of scattering may be neglected. When this occurs the wave propagation through the medium remains approximately planar, but its amplitude changes. (Even this conclusion needs to be modified if the size of the dispersed particles is too 'large').

Unfortunately, the simple situation described above is disturbed by the presence of scattering. Scattering, like absorption, may arise when the particles differ from the background material in their density, longitudinal or shear moduli, or their thermal properties. The magnitude of the scattering at a given frequency is strongly dependent on the size of the dispersed particles. However, the crucial point here is that the presence of sufficient scattering produces a distortion in the plane wave propagating through the material. In energy terms alone, absorption mainly determines how much ultrasound emerges from the far side of the material, but for high particle number densities, long path lengths, or highly scattering particles, equation (6.1) is no longer valid. Instead a contribution to the total amplitude of the wave, from the *incoherent* wave, produced by scattering, must be included. As the amplitude of the incoherent wave cannot be expressed in terms of the same exponential decay as A in equation (6.1), so neither can the total amplitude. (In fact the incoherent wave may increase in amplitude with increasing path length.) This effect is very important when determining the attenuation coefficient using the phase-sensitive detectors which are normally employed in ultrasonics. Interference between the coherent and incoherent waves means that the measured attenuation coefficient will always be an overestimate of the true attenuation in multi-phase materials (Marcus and Carstensen, 1975).

Similarly, the velocity of the ultrasonic wave will no longer be given by equation (6.2) when there is significant absorption and/or scattering. In the most favourable circumstances, when the dispersed particles are sufficiently small compared with the wavelength, so that scattering may be neglected, it is possible to use equation (6.2) to calculate the wave velocity, provided that some kind of 'effective' modulus is used. The simplest approach to this problem, originally due to Wood (1941), assumes that both the density and modulus of the multi-phase material are simply volume fraction averages of their values in the separate media. So,

$$\rho_e = \phi\rho_1 + (1-\phi)\rho_2 \tag{6.4}$$

and

$$B_e = \left(\frac{\phi}{B_1} + \frac{1-\phi}{B_2}\right)^{-1} \tag{6.5}$$

where ϕ is the volume fraction of the phase with density ρ_1 and bulk modulus B_1. Combining equations (6.4) and (6.5) leads to an expression for the velocity of sound in the composite material which is quadratic in the volume fraction ϕ over its whole range (i.e. 0 to 1; Urick, 1947)

$$v^2 = \frac{v_2^2}{\left[1-\phi\left(1-\frac{\rho_1}{\rho_2}\right)\right]\left[1-\phi\left(1-\frac{v_2^2\rho_2}{v_1^2\rho_1}\right)\right]} \tag{6.6}$$

where v_1,ρ_1 and v_2,ρ_2 are the velocities and densities of the dispersed and continuous phases, respectively.

Thus far, the spatial distribution of particles in the continuous phase has been ignored although such a distribution may only be considered to be random for $\phi \leqslant 0.01$. Allowance can be made for this, within the framework of the above approach, by introducing a parameter, α_t, known as the 'tortuosity'. α_t is a measure of the effective path length through the continuous phase relative to the direct distance between two points. Equation (6.4) is replaced by (Schwartz and Johnson, 1984)

$$\rho_e = \rho_2\left(\frac{\alpha_t\phi\rho_1 + (\alpha_t-1)(1-\phi)\rho_2}{\phi(1-\phi)\rho_1 + (\alpha_t-1+\phi^2)\rho_2}\right) \tag{6.7}$$

For $\alpha_t \gg 1$, equation (6.7) reduces to equation (6.4). For $\alpha_t = 1$, equation (6.7) becomes

$$\frac{1}{\rho_e} = \frac{\phi}{\rho_1} + \frac{(1-\phi)}{\rho_2} \tag{6.8}$$

The velocity of sound through all such effective media is given by

$$v^2 = \frac{B_e}{\rho_e} \tag{6.9}$$

There are several problems with the above approach which cannot be discussed fully in this chapter. In particular, for a solid dispersed phase, with $\phi = 1$, equation (6.6) does not reduce to equation (6.2), with M given by equation (6.3). The above theory is further unable to cope with media in which the particulate phase, instead of being dispersed, forms a separate connected phase. Completely different values of the velocity are found for two identical samples of closely packed, random glass beads, saturated with water, differing only in whether or not the beads are lightly fused together. The sample containing the solid 'frame' exhibits two different values of the speed of longitudinal sound waves, one faster and one slower than the speed of sound in the non-sintered sample, (Schwartz and Johnson, 1984). An explanation of the phenomenon is provided by Biot's theory of wave propagation in porous media (Biot, 1956).

However, it is probably its failure to take account of scattering which causes the greatest problems for effective medium theory. This theory only works well when applied to two miscible liquids at all volume fractions, or when applied to emulsions with $\lambda \rightarrow \infty$ and $\phi \rightarrow 0$.

In principle, it is possible to allow for scattering by using a so-called *multiple scattering* theory. This theory takes into account not just scattering of the incident plane wave by any given particle, but the subsequent scattering of waves which have already been scattered once. There are two difficulties with applying this theory. Firstly, there is no exact mathematical solution to this problem in the general case, for any number of scattering events. Secondly, although this is often not appreciated, the fate of the emergent ultrasonic wave is strongly dependent on all the assumptions that have been made so far, and further, on the geometry of the experimental set-up.

In the present state of the theory, a multiple scattering approach works best when the particles scatter more strongly than they absorb, and when the combined effect of particle number density, path length through the material and scattering from individual particles is small. Under these circumstances, multiple scattering theory, whose exact solution contains an infinite series, may be terminated after a small number of terms, with a corresponding restriction in the number of scattering events. As yet, even the magnitude of the second order term, for dispersed spheres, has not been generally agreed on (Javanaud and Thomas, 1988). One important feature of his theory is that the velocity and attenuation of ultrasound, in a given composite material, depend only on the product ka, where k is the wavenumber in the continuous background and a is the radius of the spherical particle.

From the above discussion, it is to be expected that very different values of ultrasonic velocity and attenuation, at a given frequency, may be obtained when either the size of the dispersed phase or its physical state changes. This effect has been studied by measuring the ultrasonic velocity and attenuation, over the frequency range 5–96 MHz, for dispersions of octadecane in water (Javanaud *et al.*, 1991). Due to the phenomenon of supercooling, it was possible to compare octadecane emulsions with octadecane suspensions at the same temperature. Thus, effects due to a change of state alone could be monitored. Furthermore, the effect of both the particle size and the size distribution on the ultrasonic properties was measured. These measurements were found to agree quite well with the multiple scattering theory.

The complexity of ultrasonic scattering does not render it a suitable subject for this chapter. However, it is worth pointing out that in the case of low concentration dispersions, scattering measurements may be of value. In principle, for a given particle size, the amplitude of the *backscattered* signal depends on the concentration of the dispersed phase. The angular distribution of the scattered energy is related to the particle size. Particularly where there is a range of sizes, the latter relationship may be very complex (Javanaud, 1989). In addition, the angular scattering measurements are difficult to make, partly due to the low intensities involved. If only relative measurements are required, the backscattered signal alone can be determined fairly simply by using the same transducer as emitter and detector. Scattering phenomena have been widely investigated in medical ultrasonics (Campbell and Waag, 1984).

6.2.4 Summary

It is clear from the above that sound propagation in simple media is well understood, but that the inclusion of structural features complicates considerably the theoretical analysis. In any real application, it may be possible to allow for many of the additional factors by comparison with a reference sample, or calibration procedures. However, it must be emphasised that these procedures should be carried out only in full knowledge of the assumptions made, to ensure that variation in ultrasonic behaviour can be attributed only to variation in the property of interest.

6.3 Measurement of ultrasonic properties

6.3.1 Introduction

As with the theory, measurement of the velocity and attenuation of ultrasound is more easily undertaken through pure materials than through mixtures. As already mentioned, the velocities of ultrasound in most liquids lie within the fairly narrow range of about 1000–2000 m s^{-1}. On the other hand, their attenuation spans a range which varies by nearly a factor of 10^3. At 1 MHz and 20°C, water and castor oil have the fairly extreme values of attenuation of 2.5×10^{-2} m^{-1} and 10 m^{-1}, respectively. However, these differences in range between velocity and attenuation are to a large extent offset by differences in the accuracy with which they may each be measured. Thus, while a precision of better than 1 part in 10^3 for velocity measurements is fairly easily achieved, the comparable figure for attenuation would be only about 10%. This chapter is concerned exclusively with techniques used for liquids, as all the inhomogeneous materials discussed here have either liquid or viscoelastic backgrounds. Thus, similar techniques tend to have been extended to these media.

It was stressed in section 6.2.1 that the discussion there applied only to the propagation of monochromatic, plane waves. The earliest ultrasonic measurements were actually carried out using continuous waves. However, most of these techniques were long ago replaced by the use of pulses (Pinkerton, 1947). The two main reasons for this are that waves which have travelled along different paths now differ in their detected arrival times, and that much smaller *mean* powers are involved. The former fact allows spurious waves (produced, for example, through unwanted reflections) to be rejected. The latter ensures that there will be negligible heating of the sample, though it is still necessary to ensure that the power within each pulse remains low enough to satisfy the criterion of the section 6.2.2. Both ultrasonic velocity and attenuation are dependent on temperature (and to a lesser extent on temperature gradients), and this needs to be controlled to within about $\pm$ 0.1°C if the precision referred to in the last paragraph is to be approached.

Despite its advantages, the pulse method must be used with care. In media with significant attenuation, the different frequency components are attenuated by different amounts, so that the pulse changes shape as it propagates. This problem can only be alleviated by using a sufficiently narrow frequency band which is equivalent to a corresponding long pulse (such pulses have come to be known as tone bursts). However, this requirement conflicts with the practical needs for measurement of the *group* velocity, which is the speed with which the wave packet, as a whole, transports energy. The group velocity may be determined by timing the passage of the pulse over a fixed known distance (Howe *et al.*, 1986). On the other hand, it is necessary to use either a continuous wave, or a tone burst if the *phase* velocity is required. The velocity at a given frequency is found by determining the distance corresponding to a given whole number of wavelengths in the liquid. This may either be measured directly (Andreae *et al.*, 1958) or deduced using interferometric techniques (Champion *et al.*, 1990).

6.3.2 *Practical considerations*

Before going into further experimental details, it is necessary to consider the practical problems which arise when the hypothetical plane wave is replaced by one produced by a real ultrasonic transducer. This is possibly the most inconvenient aspect of all the deviations from a simple theoretical model. The finite size of the ultrasonic transducer leads to diffraction of the sound wave. A clear exposition of this subject was originally given by Pinkerton (1949), although numerous more recent papers have taken it further (e.g. Gitis and Khimunin, 1969). Furthermore, even when the same transducer is used to produce waves at both its fundamental frequency and at higher harmonics, the diffraction effects (which depend on the transducer size relative to the ultrasonic wavelength) need to be evaluated separately at each frequency. The emitted sound wave is detected either by a second transducer or, after reflection, by the emitting transducer. In recent years, almost all applied work has been carried out using piezoelectric transducers to produce and detect the sound waves. These transducers are (usually) quartz crystals which are capable of converting electrical waves into sound waves and vice versa. They are *phase-sensitive* devices whose response is proportional to the instantaneous average of the complex pressure over the area of the receiver (Busse and Miller, 1981). (In this context, the term complex denotes the presence of both real and imaginary components.) This means that any deviation from a plane wave form over the transducer surface will lead to a reduced voltage output. This phenomenon, when translated into measurement terms, appears as a distortion of equation (6.1) for the exponential dependence of the attenuation. The measurement of the velocity is affected in a similar way by the deviations from planar conditions in the pressure field. It has been shown (Gitis and Khimunin, 1969) that the measured phase velocity is too high, while the group velocity is too low. These piezoelectric transducers are furthermore sharply directional devices, so that they may not be very well suited to higher frequency applications. In these circumstances it is difficult to find and/or maintain a perfectly parallel orientation between the faces of the transmitting and receiving transducers, particularly in industrial applications.

As far as the measurements of attenuation and backscattering are concerned, it is likely that far more use will be made of *phase-insensitive*, or acoustoelectric, transducers in the future (Busse and Miller, 1981). This is all the more likely since in addition to the ever present diffraction of the wave mentioned above, the passage of sound through inhomogeneous media leads to further distortions of the wave front. This phenomenon has already been mentioned in the previous section, in connection with scattering. Unlike diffraction, it is almost impossible to calculate these effects even in principle. Thus, not only are errors introduced into the measurements, but even the magnitude of any corrections is difficult to estimate. The electrical response of a phase-insensitive receiver is proportional to the instantaneous average of the *magnitude* of the complex pressure over the receiver aperture. Thus, such devices are able to make accurate direct measurements of the *power* carried by an ultrasonic wave. Busse and Miller (1981) have made a direct comparison of the phase-sensitive and phase-insensitive techniques by using a single crystal of cadmium sulphide as both a piezo- and an acoustoelectric receiver.

It often turns out to be easier to measure the non-uniformity of an ultrasonic wave than to attempt to compute it. Such measurements are nowadays performed using devices known as hydrophones. Hydrophones are phase-sensitive devices which operate over a small area, typically a few mm^2. Provided the requisite time is available, they may be used to map out the pressure field emanating from a given transducer after passage through any given material. Hydrophones are calibrated at laboratories such as the National Physical Laboratory using a laser interferometer. They are also used in some of the more sophisticated methods of measuring velocity and attenuation which employ fast Fourier transforms (FFT) of broadband pulses (Champion *et al.*, 1990). The information acquired about the phase and amplitude as functions of frequency, is converted directly into values of phase velocity and attenuation, respectively. Modern versions of this technique, for application to liquids and solids respectively, are described by Challis *et al.*, (1991) and Kline (1984).

Bearing in mind some of the shortcomings referred to above, further details will now be provided of three techniques to measure the velocity and attenuation of ultrasound in liquids and dispersions. The techniques have been in use for some time at the AFRC Institute of Food Research (IFR) at Norwich.

6.3.3 Technique to measure phase velocity and attenuation in liquids and dispersions

The first method, which measures *phase* velocity and attenuation, is a modern version of the pulse technique described by Andreae *et al.*, (1958). A schematic diagram of the equipment is given in Figure 6.1. The technique makes use of the single pass of a tone burst of the ultrasonic

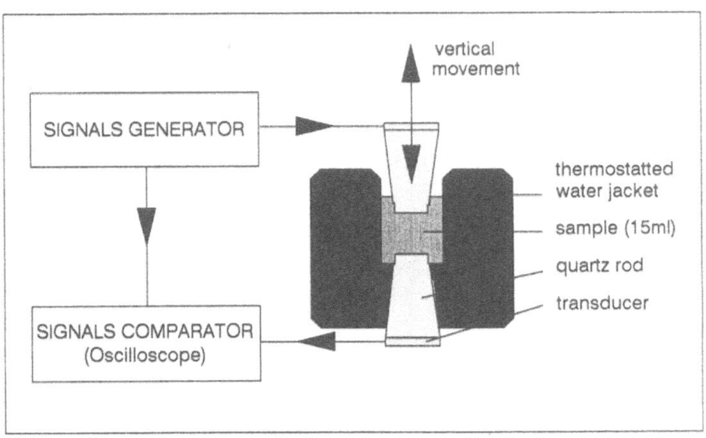

Figure 6.1 Technique to measure phase velocity and attenuation.

wave from a transmitting to a receiving quartz piezoelectric crystal, *via* quartz buffer rods and the sample liquid. The buffer rods protect the crystals from direct contact with the liquid, and the path length of the ultrasound through them must be taken into account when allowances are made for diffraction effects. The liquid is contained in a glass cell which is thermostatically controlled to within ± 0.1°C by a water bath. A minimum of 20 cycles of the ultrasonic wave is used in order to provide a well-defined amplitude and frequency for the transmitted signal.

For the velocity measurements, the sum of the input and the received signals is observed on an oscilloscope. The positions of the upper piezoelectric crystal are recorded when consecutive null signals are detected. The distance between any two nulls is equal to the wavelength of ultrasound in the liquid. Measuring the position of many consecutive nulls enables the mean wavelength to be obtained using linear regression statistics. The phase velocity of the ultrasound is the product of its wavelength and its frequency, and is thus easily calculated. There is negligible error associated with the determination of the frequency from a high quality signal generator, but it must be remembered that this figure refers to the continuous wave frequency. Thus, it is only a measure of the centre frequency of the tone burst and does not provide any information on the distribution of frequencies within the burst. This distribution has been discussed by Matsuzawa *et al.* (1987).

The attenuation of the sound wave by the liquid is determined by recording the path length in the liquid corresponding to a given amount of attenuation. When attenuation is introduced electrically into the received signal the signal amplitude will fall. The amplitude may be restored to its previous value by moving the piezoelectric transmitter closer to the receiver. Typically, an electrical attenuator will operate from 0–12 db in 2 db steps, or smaller intervals for less attenuating liquids and dispersions. The equipment in use at IFR has a quartz crystal of 16 mm diameter, with 5 MHz fundamental frequency. With the help of matching units in the electronic circuitry, the crystal may be operated on odd harmonics, up to 145 MHz in water. For more highly attenuating liquids the maximum attainable frequency is much lower, due to loss of signal. For water, although the velocity may be measured over the whole frequency range, the attenuation is too low to be measured at frequencies below 45 MHz. Before carrying out any measurements, it is essential to ensure optimum alignment of the transducers, by adjusting the angle of the face of the lower transducer, at the highest frequency at which measurements are going to be made.

Under the best circumstances, the precision of the above equipment in measurement of velocity is a few parts in 10^4 or approaching 0.3 m s^{-1}. Greater precision than this would necessitate better temperature control, as a temperature change of 0.1°C, near room temperature, typically changes the velocity of sound in liquids by about 0.2 m s^{-1}. The errors due to diffraction effects are also of this order of magnitude (Gitis and Khimunin, 1969), and as already mentioned, are very difficult to estimate. When the equipment is used with dispersions, rather than with pure liquids, the further deviation of the ultrasonic wave from planar will cause an increase in the uncertainty associated with the velocity. Low volume fraction dispersions cause less perturbation of the wave front than higher volume fractions if all the other parameters are unchanged.

For attenuation measurements the errors vary considerably, with the best results obtained from the middle of the range of attenuations. Liquids with low attenuation necessitate long path lengths for accurate measurements, and are subject to diffraction errors at the lower frequencies. For very highly attenuating liquids, it is the measurement of distance which contributes most to the uncertainty in attenuation. In all cases, the attenuators themselves, and the visual method using the oscilloscope are substantial sources of errors. (Work is in progress to measure the velocity and attenuation automatically on a modified form of this equipment.) The set-up described here has an advantage that attenuation may be measured over a range of frequencies with the same equipment. As with velocity, the accuracy of the measurements is reduced if dispersions are being used instead of liquids. Unfortunately, for dispersions, the achievement of an easily measured attenuation occurs in the regime of high acoustic depth where single scattering theory or first-order multiple scattering no longer apply.

6.3.4 Double-liquid technique for measurement of velocity difference

It is possible to use similar equipment to that just described to measure the *differences* in velocity and attenuation between a standard liquid and the unknown sample (Carstensen, 1954). The two liquids are separated by a Mylar screen, with one transducer in each liquid. Distilled water, with its well known value of velocity, and its low attenuation, is usually the most suitable reference liquid. The separation between nulls is now governed by the difference in the wavelengths of sound in the two liquids, so that greater accuracy is achieved. Furthermore, the relative path length in the different liquids is altered by moving both transducers together so that the distance between them remains constant (within a given set of measurements). This greatly reduces any errors caused by diffraction. When the two liquids are similar, e.g. water and a dilute aqueous dispersion, errors due to temperature variation are also minimised. Although this technique is capable of greater accuracy, particularly at low frequencies, it has the drawback that in general larger quantities of the sample liquids, (or dispersions) are required. Nevertheless, a modern version of this equipment was used at IFR to measure the velocity of sound in a series of vegetable oils. These measurements were then related to a simple theory in which the velocity of ultrasound in the oils depends on their fatty acid composition (Javanaud and Rahalkar, 1988).

6.3.5 *Measurement of group velocity in liquids and dispersions*

Another method currently in use at the Institute of Food Research measures the *group* velocity of ultrasound in liquids and dispersions *via* a pulse time-of-flight technique (Howe *et al.*, 1986). A schematic diagram of the equipment is given in Figure 6.2. A water bath, thermostatically controlled to $\pm$ 0.1°C, contains the quartz transducers (probes) facing one another horizontally. Between the transducers lies a perspex cell containing the sample liquid. Measurements may be made at different heights in order to monitor sedimentation processes in dispersions.

In this pulse method, the emitting transducer is excited by a single high voltage pulse of duration $\leqslant$ 100 ns. As this is less than the natural period of the sender, ($\simeq$ 150 ns for the 6.4 MHz transducer normally used), the shape of the wave packet is determined by the electromechanical properties of the emitter, not by the pulse shape. The short duration of the pulse permits an accurate determination of its transit time using a pulse-overlap method. The emitted and received signals are observed on an oscilloscope and the repeat rate for the transmitted pulse is varied until the received signal overlaps successive transmitted pulses. At this point the reciprocal of

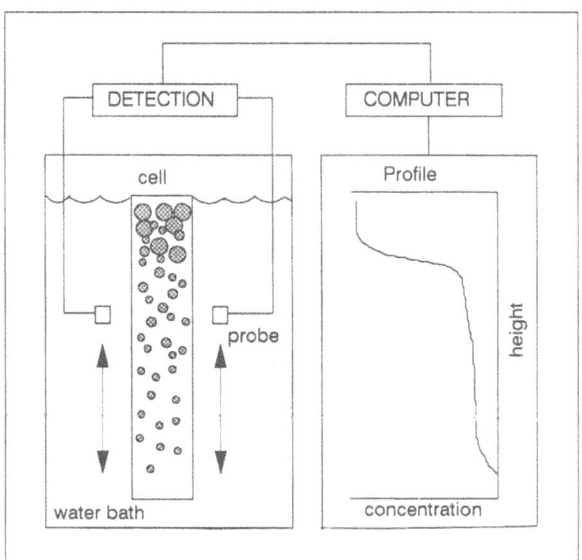

Figure 6.2 Technique to measure group velocity in dispersions as a function of height.

the pulse repetition rate equals the total travel time between the transducers. The total travel time depends on the group velocity of sound in the liquid (and on the path length through the liquid and an unknown constant, Wilson *et al.*, 1991). By calibrating the cell with two liquids which possess known values of the velocity of ultrasound, the velocity of ultrasound in the sample may be determined. Perfect overlap of the transmitted and the received wave packets with each other is impossible, due to the absorption of sound by the sample liquid. The error caused by the difference in shape between the two pulses depends on the amount of attenuation which the received pulse has undergone. In practice, this leads to an error in the velocity measurement of typically about 0.05% (Champion *et al.*, 1990).

6.3.6 Summary of techniques available

In an overview, it is not possible to describe in detail the full range of techniques that could be applied to food materials. We have expanded on two types of measurement, namely: phase velocity and attenuation over a range of frequencies, and a pulse method for determining group velocity, particularly as a function of position. Other measurements that can be made include broadband Fourier methods, currently being developed to a high specification by Challis *et al.*, (1991). Imaging methods are widely used in medical work (Sinclair *et al.*, 1982) and could be adapted for use on foods. In addition, mention should be made of the use of Doppler shift techniques to measure the flow of fluid in pipes (Lynworth, 1979; Sanderson, 1982).

6.4 Applications to food systems

6.4.1 Overview of methods used on foods and related materials

The study or monitoring of food systems using ultrasonic techniques is made advantageous by the fact that they are non-invasive, reasonably accurate and relatively cheap. Furthermore sound waves travel easily through many materials that are opaque to light. They are, however, strongly attenuated by gases. This means that it is difficult to apply ultrasonic techniques to media with a large air content.

Of the experimental methods described in the previous section, the one which has been used most with foods is pulse overlap (or the similar pulse-echo overlap) to measure the velocity of sound. This method has the great advantage of requiring only a fixed path length for the sound, so that it is more easily adapted to on-line situations. The time needed to make the measurements required to determine the velocity of sound is also generally less than that required in variable path length techniques.

In the last few years several reviews have appeared on the subject of ultrasonic applications to food. Those by Povey and McClements (1988) and Povey (1989) form one paper which is split into two parts. The first part deals with theory and experimental techniques, while the second is devoted to applications. Each part is longer, and therefore considerably more detailed, than the review by Javanaud (1988). Furthermore, Povey and McClements (1988) provide physical data on sound propagation in degassed and distilled water, while Povey (1989) provides similar data for a range of foods and food-related materials. These data are not included in Javanaud's review which summarises previous work, and then discusses its limitations and future developments.

Table 6.1 shows an overview of the ultrasonic measurements that have been carried out on foods. The methods will be discussed in turn.

6.4.2 Velocity measurements

Most of the velocity measurements have been carried out with a view to measuring the concentration of a minority component in a mixture. Often the component monitored is present as solid particles dispersed in a liquid or viscoelastic background. This approach has been applied to the first seven materials listed in Table 6.1. However, many previous investigators were unaware of the relevant theory (presented in section 6.2). Thus, many of them have assumed, sometimes wrongly, that there is a linear relationship between the concentration of the dispersed phase and its contribution to the velocity of sound in the systems of interest. Others have ignored theory completely and tried to proceed by establishing an empirical relationship between these quantities. An exception is the work at the Institute of Food Research on emulsions by Howe *et al.*, (1986). They were able to monitor the creaming of droplets in an emulsion, with time, by taking measurements of the velocity of ultrasound at different heights. Under a wide range of circumstances Urick's equation (equation (6.6) in section 6.2.3) was applicable.

One of the exceptions to the above use of velocity measurements occurs in meat, where the background material is a solid, rather than a liquid. This, however, amounts to a difference of detail rather than principle. This is one of the few applications which has been used extensively, to determine the fat content of live animals (Miles and Fursey, 1976). In this application, as with other solid samples, either a soft-tipped probe or an acoustic couplant, such as paraffin, needs to be used to avoid transmitting the ultrasound across an air gap.

Another use of velocity measurements has been to detect the ageing process in fruit and eggs. So far, no consensus of opinion on this subject has emerged (Javanaud, 1988). It should be borne in mind that in nearly all

Table 6.1 Ultrasonic methods used to evaluate foods

Velocity		
Percentage of minority phase		
Fruit juices	Fruit flesh, sugar	Zacharias and Parnell (1972)
Wine	Solids and alcohol	Winder *et al.* (1970)
Milk	Solids and fat	Fitzgerald and Winder (1961); Miles *et al.* (1990)
Yeast slurry	Solids	Feil and Zacharias (1971)
Oils	Solid fat	McClements and Povey (1987)
	Adulteration	Rao *et al.* (1980)
Emulsions	Oil	Howe *et al.* (1986); McClements *et al.* (1990)
Ice cream	Water	Steele (1964)
Meat	Solid fat	Miles and Fursey (1976)
Age		
Fruit		Garrett and Furry (1971); Hayes and Chignon (1982)
Eggs (white and yolk)		Mayer and Hiedemann (1959); Povey and Wilkinson (1980); Javanaud *et al.* (1984)
Structure or texture		
Ice cream		Smith and Wittinger (1986)
Biscuits		Povey and Harden (1981)
Phase changes		
Oils		McClements and Povey (1988)
Emulsions		Dickinson *et al.* (1990); Javanaud *et al.* (1991)
Water/ice in meat		Miles and Cutting (1974)
Attenuation		
Percentage of minority component		
Milk and cream	Fats, proteins	Miles *et al.* (1990)
Stability		
Orange juice		Sarkar and Wolfe (1983)
Age		
Fruit		Sarkar and Wolfe (1983)
Potatoes		Sarkar and Wolfe (1983)
Eggs (white and yolk)		Povey and Wilkinson (1980); Javanaud *et al.* (1984); Choi *et al.* (1987)
Phase change		
Emulsions		Javanaud *et al.* (1991)
Ice/water mixture in meat		Miles and Shore (1978)
Scattering		
Fish	Lipid content	Freese and Hamid (1974)
Beef	Quality grade	Haumschild and Carlson (1983)
Imaging		
Fish	Presence of sealworms	Hafsteinsson *et al.* (1989)

the applications discussed here, the temperature of the material would not be controlled to within ± 0.1°C, so that the accuracy of the measurement does not match the optimum values given in section 6.3.3. For biscuits and ice cream, ultrasound was used to probe the 'structure' of the material. From equation (6.2), it should be noted that such changes will only be detectable if they alter the modulus. In the case of biscuits a correlation was found between crispness and velocity of sound (Povey and Harden, 1981). The crispness was assessed by a panel of people. In ice cream the ultrasonic velocity measurements were found to correlate with such structural parameters of the mixture as viscosity, water binding and the energy required for freezing to occur (Smith and Wittinger, 1986).

An interferometric technique for measuring the velocity of ultrasound has been used to detect the adulteration of fats and oils (Rao *et al.*, 1980). The sensitivity of this method, which depends on the difference between the velocity of ultrasound in the adulterant and in the pure oil, could be improved if the double-liquid system, described in section 6.3.4, were to be employed. In this case, it would be desirable to put the adulterated liquid and the corresponding pure oil on either side of the Mylar screen. The null separation varies inversely with velocity difference, so this method is particularly sensitive when very similar liquids are compared.

The use of ultrasound to detect phase changes in one component of emulsions has already been mentioned in section 6.2.3 (Javanaud *et al.*, 1991). Other work on oil mixtures alone (McClements and Povey, 1988) also detected significant changes in velocity as oil solidifies or melts. McClements and Povey studied 15% w/w triglyceride/paraffin oil mixtures over a wide range of temperature (0–70°C). They found that Wood's equation gave a fairly accurate means of predicting the solid oil content from the velocity measurements. This is to be contrasted with the work of Javanaud *et al.* (1991), where due to the much greater difference in physical properties between the two phases, a more sophisticated theory was needed to explain the results. They also showed that the measurements depend on both the mean size of the dispersed phase and on its size distribution. The pronounced effects observed only occur when the ultrasound undergoes a resonance phenomenon. This happens at certain frequencies if the solid particle size is such that $ka \geqslant 1$ (see section 6.2.3). In most applications the particles are not the perfect spheres assumed in the theoretical treatment. Nevertheless, it might be possible to exploit the resonance phenomenon in real foods. Miles and Cutting (1974) have found that as the temperature of meat falls from 0 to –5°C, the velocity of ultrasound increases by 70%. The exact magnitude of this change would be very difficult to predict theoretically because of the complex structure of the water in meat.

6.4.3 *Attenuation measurements*

Until now, sound attenuation measurements have been used considerably less than velocity measurements with real foods. There are two main reasons for this. The first is the practical consideration, already mentioned, that it is not easy to provide the variable path length necessary for accurate attenuation measurements. The second reason is the fact that while it is straightforward to apply an (often erroneous) linear relation, or the quadratic Wood's equation to determine the proportion of a minority component from velocity measurements, there is no correspondingly simplistic interpretation for the attenuation.

The work carried out by Sarkar and Wolfe (1983) tried to overcome the variable transducer separation problem by successively immersing two samples, differing only in their thickness, in water. With this method, significant errors are caused by the difficulty in obtaining reproducible sample and experimental conditions. It had the further disadvantage that, for the foods which they studied, the attenuation coefficient changed with time. Water gradually entered the intercellular void spaces of the fruit and vegetables, causing large changes in the propagation of ultrasound. Unlike liquid or viscoelastic materials, solid materials shaped like vegetables and fruit pose a particularly tricky problem as far as the use of ultrasound is concerned. It might be better to apply the method used by Shore *et al.* (1986) for bovine skeletal muscle. The attenuation of the specimen, which is sandwiched between buffer rods when making the measurements, is compared with that of water of the same total path length. Considerable care needs to be made to correct for reflections. The drawback of this technique for fruit and vegetables is that their porous nature, and hence high attenuation, necessitates low path lengths. This leads to loss of accuracy in reproducing the distance for comparative measurements and in measuring the distance for the absolute determination of attenuation.

The work on the attenuation of sound by egg white by Javanaud *et al.* (1984) was confirmed by Choi *et al.* (1987). Neither group detected an ageing effect. An effect was expected (Povey and Wilkinson, 1980) since the relative proportion of thick and thin white changes with time and the former has a higher viscosity and a different structural organisation from the latter. This use of ultrasound resembles that employed to monitor the structure of ice cream (Smith and Wittinger, 1986) or biscuits (Povey and Harden, 1981).

The attenuation of ultrasound through emulsions or meat is very sensitive to phase changes. For example, in a 5% v/v octadecane emulsion in water, the magnitude of the attenuation nearly doubles when the octadecane droplets solidify (Javanaud *et al.*, 1991). In meat, Miles and Cutting (1974) found that the attenuation increased by an order of magnitude while the phase change was occurring. (Unlike the emulsion system, the attenuation falls again at lower temperatures as the reason for the high attenuation in this case differs from that in suspensions.) This application is clearly suited to on-line measurements, provided that a suitable experimental set-up can be developed.

For both dairy and alkane emulsions, it is in principle possible to deduce the size distribution of the droplets as well as their concentration (Miles *et al.*, 1990, Javanaud *et al.*, 1991). However, such measurements necessitate a prior knowledge of the material composition and the use of a broad range of frequencies. Fourier transform techniques (e.g. Challis *et al.*, 1991) would be useful here as the attainment of a wide frequency range through spot frequency measurements is very time consuming.

6.4.4 Other ultrasonic applications

For completeness, other applications of ultrasound in the food industry have been included in Table 6.1. Possibly as a result of the widespread use of sonar to detect shoals of fish (e.g. Javanaud, 1992), considerable research has been directed towards fish. The ultrasonic properties of freshwater fish were studied some time ago by Freese and Makow (1968). More recent developments in ultrasonic techniques include Scanning Acoustic Microscopy (SAM) imaging (Sinclair *et al.*, 1982). A modified imaging technique, Scanning Laser Acoustic Microscopy (SLAM) has been applied to the detection of sealworms in fish tissue (Hafsteinsson *et al.*, 1989).

The use of high-power, low-frequency ultrasound to control chemical reactions (sonochemistry) has so far been ignored in this chapter. Until recently, there have been few applications to food systems. However, lately there has been some interest, in particular for controlling crystallisation processes, for example in ice lollies or in chocolate (Sullivan, 1992). Another reported use is to destroy the tough connective tissue in raw meat, tenderising the fibre without a lengthy hanging period (Sullivan, 1992).

6.5 Future developments

Table 6.1 shows that, to date, there have been many attempts to use ultrasound to monitor various properties of food systems. The most successful application, so far, has been the measurement of the velocity of ultrasound to determine the concentration of a dispersed (liquid or solid) phase in a liquid, or viscoelastic background material.

Future progress depends to some extent on whether the absolute value of the volume fraction or only its change with time is sought. If absolute values are required, the equipment must be calibrated using a standard liquid, of similar acoustic impedance to the sample and possibly also with a range of dispersions with variable volume fraction of the dispersed phase. (These dispersions would need to have the same size distribution as that of the sample.) The latter procedure is essential if the theory is not easily applied, and may enable an empirical approach to be taken.

To obtain reliable results, it is furthermore necessary either to know the temperature of the material, or to stabilise it. The desired degree of temperature control may vary considerably, being determined by both the particular system being monitored and by the required degree of accuracy. It was suggested some time ago (Steele, 1964) that automatic temperature compensation is desirable. If the temperature cannot be controlled, then, provided it can be measured and does not change too rapidly, fruitful velocity measurements may still be made. It is then necessary to carry out any calibration procedures over a range of temperatures. Depending on the degree of technical sophistication of the measuring equipment this may not be such a drawback.

In the near future, particularly when the change in the velocity of ultrasound is the measured quantity, it seems likely that the pulse-overlap (and pulse-echo overlap) method will continue to have widespread application. It should be stressed again that this method, which measures the group velocity of sound, has the practical advantage of using a fixed path length through the food material. In principle, interferometric techniques for measuring phase velocity may be operated with a fixed path length, using a variable frequency. However, Povey and McClements (1988) found it difficult to prepare solid food samples with surfaces which were sufficiently smooth and parallel that the standing wave resonances were observable. This problem would also be present in many cases where food is flowing along solid pipes or is enclosed by a solid chamber. Such containers are rarely designed to meet the criteria for a good interferometric cell.

It is to be hoped that the main thrust forward with the pulse-overlap technique in the next few years will be an extension of the range of foods to which it is applied. All aspects of these applications will benefit from further simultaneous experimental and theoretical work on similar, but simpler, model food systems.

At present, use of the more accurate attenuation measurements described in Section 6.3 is likely to be restricted to laboratory work on model food systems. It is possible to make on-line measurements with foods, using a fixed path length, by comparing the amplitude of the received signal with that obtained using a comparison liquid of similar acoustic impedance. Such measurements will, however, be fairly crude, even if the calibration liquid also has a similar attenuation to the sample material. It seems likely that in the near future such techniques will be restricted to detecting *changes* in the sample attenuation. In particular, phenomena such as solidification or melting of the dispersed phase should be easily detectable in this way. This assumes that there is a negligible change in other physical parameters which affect the attenuation measurements. It is unlikely that this kind of measurement will be simply related to theory, except possibly for simple systems containing very low volume fractions of the dispersed phase (Javanaud *et al.*, 1986) due to the nature of the problems discussed in sections 6.2 and 6.3. It seems much more probable that future research will be ultimately directed towards the use of phase-insensitive receivers (Busse and Miller, 1981). Although work needs to be done to achieve simple practical applications, it is certain that measurements from these devices would be much simpler to interpret and more easily reproducible under slightly different experimental conditions. From a practical point of view this would lead to greater flexibility in their use.

In principle, scattering measurements could provide valuable supplementary information on foods. In particular, in long path length, highly attenuating materials, the backscattered sound would be easier to detect than that transmitted through the system. Once again, it might initially be necessary to take an empirical approach. The exact theoretical determination of the backscattered wave is rather complicated, particularly for highly attenuating materials. It depends on the geometry of the system as well as on the frequency and the kind of detector employed. Certain aspects of these problems have been discussed by Javanaud (1992) and by Ishimaru (1978). It seems less likely that measurements of the full angular distribution of the scattered sound will be made, for practical reasons. As mentioned in section 6.2.3, in principle, angular scattering measurements can provide information on the size of the dispersed phase while the backscattered signal, (for a given size), is related to the volume fraction.

In view of the increasing use of ultrasonics to evaluate or monitor materials in many industrial and medical areas, there is little doubt that its application to foods will also increase. The principles and experimental techniques appropriate to foods are common to other liquid or semi-solid materials, and there is much to be gained from following closely the exciting developments in, for example, dispersion ultrasonics (Challis *et al.*, work in progress) or sonochemistry (Mason and Lorimer, 1989). With such a rich literature and considerable on-going work in several related fields, we are optimistic that the potential for ultrasonic methods in the food industry will be realised.

References

Andreae, J.J., Bass, R., Heasell, E.L. and Lamb, J. (1958) Pulse techniques for measuring ultrasonic absorption in liquids. *Acustica*, **8**, 131–142.
Anon (1988) Underwater acoustics. *Acoustics Bulletin*, **13**, 9–33.
Biot, M.A. (1956) Theory of propagation of elastic waves in a fluid-saturated porous solid.

I. Low frequency range. *J. Acoustic. Soc. Am.* **28**, 168–178; II. Higher frequency range. *J. Acoustic. Soc. Am.*, **28**, 179–191.
Busse, L.J. and Miller, J.G. (1981) Detection of spatially nonuniform ultrasonic radiation with phase sensitive (piezo-electric) and phase insensitive (acoustoelectric) receivers. *J. Acoustic. Soc. Am.*, **70**, 1377–1386.
Campbell, J.A. and Waag, R,C. (1984) Ultrasonic scattering properties of three random media with implications for tissue characterisation. *J. Acoustic. Soc. Am.*, **75**, 1879–1886.
Carstensen, E.L. (1954) Measurement of dispersion of velocity of sound in liquids. *J. Acoustic. Soc. Am.*, **26**(5), 858–861.
Challis, R.E., Harrison, J.A., Holmes, A.K. and Cocker, R.P. (1991) A wide bandwidth spectrometer for rapid ultrasonic absorption measurements in liquids. *J. Acoustic. Soc. Am.*, **90**, 730–740.
Champion, J.V., Langton, C.M., Meeten, G.H. and Sherman, N.E. (1990) Near field ultrasonic measurement apparatus for fluids. *Meas. Sci. Technol.*, **1**, 786–792.
Choi, P.K., Bae, J.R. and Takagi, K. (1987) Frequency dependence of ultrasonic absorption in egg white. *J. Acoustic. Soc. Am.*, **80**(6), 1844–1846.
Coulson, C.A. (1955) *Waves*. Chapter 6. Oliver and Boyd, Edinburgh.
Dickinson E., Goller M.I., McClements D.J., Peasgood S. and Povey M.J.W. (1990) Ultrasonic monitoring of crystallisation in an oil-in-water emulsion. *J. Chem. Soc., Faraday Trans.* **86**(7), 1147–1148.
Feil, M.F. and Zacharias, E.M. (1971) The determination of yeast slurry consistency and wort plato by sonic solution analysis. *Brewers Dig.*, **46**, 76–80.
Fitzgerald, J.W. and Winder, W.C. (1961) An ultrasonic method for measurement of solids-non-fat and milk fat in fluid milk. *J. Dairy Sci.*, **44**, 1165.
Freese, M. and Hamid, M.A.K. (1974) Lipid content determination in whole fish using ultrasonic pulse backscatter. *Proc. Ultrasonics Symp. IEEE*, 69–74.
Freese, M. and Makow, D. (1968) High frequency ultrasonic properties of freshwater fish tissue. *J. Acoustic. Soc. Am.*, **44**, 1282–1289.
Garrett, R.E. and Furry, R.B. (1971) Velocity of sonic pulses in apples, *American Society of Agricultural Engineering*, Paper no. 71–331, Pullman, Washington, USA.
Gitis, M. B. and Khimunin, A. S. (1969) Diffraction effects in ultrasonic measurements. *Sov. Phys. Acoustics*, **14**, 413–431.
Hafsteinsson, H., Parkin, K., Chivers, R. and Syed, S.H.R. (1989) The application of ultrasonic waves to detect sealworms in fish tissue. *J. Food Sci.*, **54**, 244–247; 273.
Haumschild, D.J. and Carlson, D.L. (1983) An ultrasonic Bragg scattering technique for characterisation of marbling in beef. *Ultrasonics*, **21**, 226–233.
Hayes, C.F. and Chignon, H.T.G. (1982) Acoustic properties of papaya. *J. Texture Studies*, **13**, 397–402.
Hill, C.R. and ter Haar, G. (1982) Ultrasound. *Non-ionising Radiation Protection*, Chapter 6. WHO Regional Publications, Copenhagen.
Howe. A. M., Mackie, A. R. and Robins, M. M. (1986) Technique to measure emulsion creaming by velocity of ultrasound. *J. Dispersion Sci. Tech.*, **7**(2), 231–243.
Ishimaru, A. (1978) *Wave Propagation and Scattering in Random Media*, Volumes I and II. Academic Press, New York.
Javanaud, C. (1988) Applications of ultrasound to food systems. *Ultrasonics*, **26**, 117–123.
Javanaud, C. (1989) The application of a fractal model to the scattering of ultrasound in biological media. *J. Acoustic. Soc. Am.*, **86**, 493–496.
Javanaud, C. (1992) Two aspects of the use of acoustics in the quantitative estimation of fish stocks. *Developments in Acoustics and Ultrasonics*. Institute of Physics.
Javanaud, C., and Rahalkar, R.R. (1988) Velocity of sound in vegetable oils. *Fat Sci. Technol.*, **90**, 73–75.
Javanaud, C. and Thomas, A. (1988) Multiple scattering using the Foldy-Twersky integral equation. *Ultrasonics*, **26**, 341–343.
Javanaud, C., Rahalkar, R.R. and Richmond, P. (1984) Measurement of the speed and attenuation of ultrasound in egg white and egg yolk. *J. Acoustic. Soc. Am.*, **76**(3), 670–675.
Javanaud, C., Lond, P. and Rahalkar, R.R. (1986) Evidence for sound absorption in emulsions due to differing thermal properties of the two phases. *Ultrasonics*, **24**, 137–141.
Javanaud, C., Gladwell, N.R., Gouldby, S.J., Hibberd, D.J., Thomas, A. and Robins, M.M.

(1991) Experimental and theoretical values of the ultrasonic properties of dispersions: effect of particle state and size distribution. *Ultrasonics*, **29**, 331–337.

Kline, R.A. (1984) Measurement of attenuation and dispersion using an ultrasonic spectroscopy technique. *J. Acoustic. Soc. Am.*, **76**, 498–504.

Lynworth, L.C. (1979) Ultrasonic flowmeter. In *Physical Acoustics*, Volume XIV (ed. Mason, W.P. and Thruston R.N.). Academic Press, New York, pp. 408–516.

Marcus, P. W. and Carstensen, E. L. (1975) Problems with absorption measurements of inhomogeneous solids. *J. Acoustic. Soc. Am.*, **58**, 1334–1335 (L).

Mason, T.J. and Lorimer, J.P. (1989) *Sonochemistry: Theory, Applications and Uses of Ultrasound in Chemistry*. Ellis Horwood, Chichester, UK.

Matsuzawa, K., Inoue, N. and Hagegawa, T. (1987) A new simple method of ultrasonic velocity and attenuation measurement in a high absorption liquid. *J. Acoustic. Soc. Am.*, **81** 947–951.

Mayer, W.G. and Hiedemann, E.A. (1959) On the feasibility of ultrasonic grading of shell eggs. *Food Res.*, **24**(1), 97–103.

McClements, D.J. and Povey, M.J.W. (1987) Ultrasonic velocity: a new method for determining solid fat contents. *J. Food. Technol.*, **22**, 419–427.

McClements, D.J. and Povey, M.J.W. (1988) Investigation of phase transitions in glyceride/paraffin oil mixtures using ultrasonic velocity measurements. *J. Am. Oil Chem. Soc.*, **65**, 1791–1795.

McClements, D.J., Povey, M.J.W., Jury, M. and Betsanis, E. (1990) Ultrasonic characterisation of a food emulsion. *Ultrasonics*, **28**, 266–272.

Miles, C.A. and Cutting, C.L. (1974) Technical note: changes in the velocity of ultrasound in meat during freezing. *J. Food. Technol.*, **9**, 119–122.

Miles, C.A. and Fursey, G.A.J. (1976) Measurement of the fat content of meat using ultrasonic waves. *Food. Chem.*, **2**, 107–118.

Miles, C.A. and Shore, D. (1978) Changes in the attenuation of ultrasound in meat during freezing. *Proc. 24th EMMRW*, Kulmbach, **1** D4.

Miles, C.A., Shore, D. and Langley, K.R. (1990) Attenuation of ultrasound in milks and creams. *Ultrasonics*, **28**, 394–400.

Pinkerton, J.M.M. (1947) A pulse method for the measurement of ultrasonic absorption in liquids: results for water. *Nature (London)*, **160**, 128–129.

Pinkerton, J.M.M. (1949) On the pulse method of measuring ultrasonic absorption in liquids. *Proc. Phys. Soc. (London)*, **62B**, 286–299.

Povey, M.J.W. (1989) Ultrasonics in food engineering. Part II Applications. *J. Food Eng.*, **9**, 1–20.

Povey, M.J.W. and Harden, C.A. (1981) An application of the ultrasonic pulse echo technique to the measurement of crispness of biscuits. *J. Food. Technol.*, **16**, 167–175.

Povey, M.J.W. and McClements, D.J. (1988) Ultrasonics in food engineering. Part I. Introduction and experimental methods. *J. Food Eng.*, **8**, 217–245.

Povey, M.J.W. and Wilkinson, J.M. (1980) Application of ultrasonic pulse-echo techniques to egg albumen quality testing: a preliminary report. *Br. Poultry Sci.*, **21**, 489–495.

Rao, C.R., Reddy, L.C.S. and Prabhu, C.A.R. (1980) Study of adulteration in fats and oils by an ultrasonic method. *Current Sci.*, **49**, 185–186.

Sanderson, M.L. (1982) Electromagnetic and ultrasonic flowmeters: their present states and future possibilities. *Electron. Power*, February, 161–4.

Sarkar, N. and Wolfe, R.R. (1983) Potential of ultrasonic measurements in food quality evaluation. *Trans. ASAE*, **26**(2), 624–629.

Schwartz, L. M. and Johnson, D. L. (1984) Long wavelength acoustic propagation in ordered and disordered suspensions. *Phys. Rev.*, **B30**, 4302–4313.

Shore, N., Woods, M.O. and Miles, C.A. (1986) Attenuation of ultrasound in post rigor bovine skeletal muscle. *Ultrasonics*, **24**, 81–87.

Sinclair, D.A., Smith, I.R. and Wickramsinghe, H.K. (1982) Recent developments in scanning acoustic microscopy. *Radio Electron. Eng.*, **52**, 479–493.

Smith, D.E. and Wittinger, S.A. (1986) Effect of sweeteners and stabilisers on the structure of ice cream mix as determined by acoustic methods. *J Food Proc. Pres.*, **10**, 227–235.

Steele, D.J. (1964) Ultrasonics to measure the moisture content of food products. *Br. J. NDT*, **16**, 169–173.

Sullivan, C. (1992) A sound investment. *Chem. Ind.*, **10**, 365.

Urick, R.J. (1947) A sound velocity method for determining the compressibility of finely divided substances. *J. Appl. Phys.*, **18**, 983–987.

Wells, P.N.T. (1977) *Biomedical Ultrasonics*, Academic Press, London.

Wilson, P.D.G., Gladwell, N.R., Hibberd, D.J. and Robins, M.M. (1991) Rationalisation of errors from two techniques measuring the velocity of ultrasound in liquids. *Ultrasonics.*, **29**, 225–229.

Winder, W.C., Aulik, D.J. and Rice, A.C. (1970) An ultrasonic method for direct and simultaneous determination of alcohol and extract content of wines. *Am. J. Enol. Vitic.*, **21**, 1–11.

Wood, A.B. (1941) *A Textbook of Sound.* G Bell and Sons, London, p. 361.

Zacharias, E.M. and Parnell, R.A. (1972) Measuring the solids content of foods by sound velocimetry. *Food Technol.*, **26**, 160–166.

7 Mass spectrometry

D. LLOYD

7.1 Introduction: basic principles

Mass spectrometry is the most powerful technique currently available for the analysis of chemical species (Beynon, 1981). Analyses are achieved by ionization and subsequent interactions of the charged species thus generated with electrical or magnetic fields; separations are thus based on mass-to-charge ratios. In order to prevent reaction with neutral species, the entire sequence of operations must be conducted in the vapour phase under high vacuum conditions. Mass spectrometry differs from other spectroscopic techniques in that it measures mass rather than frequency. First used by J. J. Thomson in 1912, and developed by Aston in 1919 into a more generally useful technique, it now finds wide application in both the physical and biological sciences on account of its high sensitivity, specificity, and information content.

All instruments employ four stages (Figure 7.1). The sample is introduced into the high vacuum, where it must be ionized before separation of the constituent ions can be achieved. Finally a current-generating device is used to collect and detect the ions. The great variety and versatility of mass spectrometry systems comes about as a consequence of

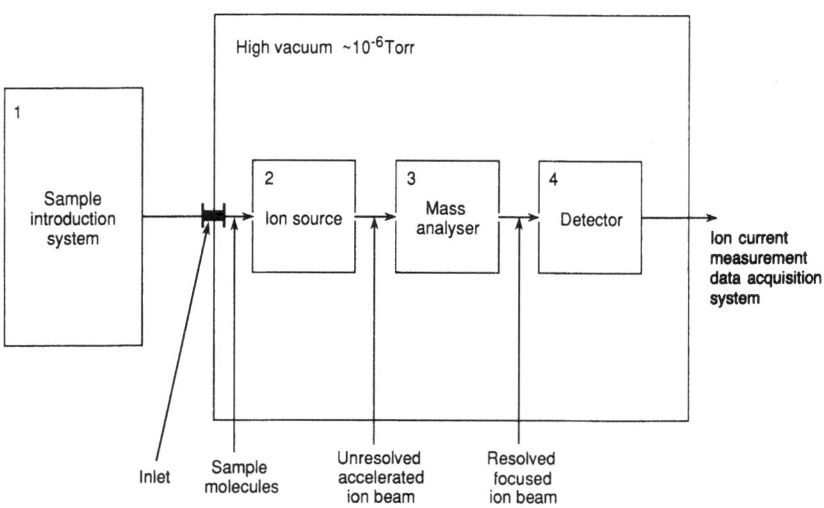

Figure 7.1 Block diagram of the elements of a mass spectrometer.

the multiplicity of available approaches to problems that arise at each stage of the process. Factors influencing the choice of the most appropriate mass spectrometric method to suit particular needs will be emphasized in this chapter.

7.2 Basic definitions

Specification of the performance characteristics of a mass spectrometer invariably include the following definitions:

Resolution. Resolving power is a measure of the ability of the mass spectrometer to separate adjoining masses (Figure 7.2). Resolution is usually a function of M.

Sensitivity. This is dependent on inlet system and on the nature of the compound under study. A standard compound (e.g. methyl stearate) is examined under standard operating conditions, and sensitivity is then

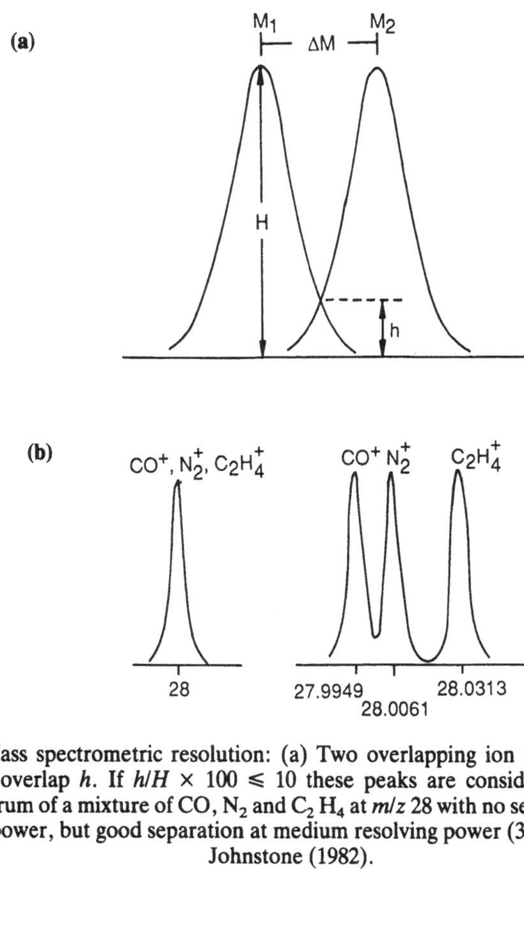

Figure 7.2 Mass spectrometric resolution: (a) Two overlapping ion peaks M_1 and M_2 of height H and overlap h. If $h/H \times 100 \leq 10$ these peaks are considered to be resolved. (b) Mass spectrum of a mixture of CO, N_2 and C_2H_4 at m/z 28 with no separation of masses at low resolving power, but good separation at medium resolving power (3000). From Rose and Johnstone (1982).

measured as coulombs μg^{-1}, or as the signal/noise ratio, for the molecular ion. The full mass spectrum of an organic compound can currently be obtained with about 1 ng of material, and target compounds can often be detected specifically at sub-picogram levels, using the technique of selected ion monitoring.

Dynamic range. This characteristic is most important for quantitative applications, e.g. in isotope ratio measurements. Specialized ion counting techniques enable isotope abundance ratios of 10^8 to be measured; speed and precision of measurement are mutually exclusive. Precise analysis of the stable isotopes of H, C, N, O and S at natural abundances are providing important ways of studying environmental problems, food additives and metabolic research (Koppenaal, 1988; Silfer *et al.*, 1991; Prosser *et al.*, 1991; Chace and Abramson, 1989). Various types of isotope ratio mass spectrometry (IRMS) include automated nitrogen and carbon analyser-MS (ANCA-MS), gas chromatography combustion-MS (GC-CMS) and chemical reaction interface MS (CRIMS).

Mass accuracy. This is determined by dispersion, stability of electronics and abundance sensitivity; known mass standards (e.g. xenon gas) are used as references, and high performance instruments have mass accuracies of the order of 1 ppm.

7.3 Ion analysis

7.3.1 Magnetic sector mass spectrometers

In a double-focusing analyser, an electric sector focuses ions according to their translational energies before they are passed through a magnetic field, where separation on the basis of mass-to-charge ratio is achieved. In this arrangement (Figure 7.3)

$$m/z = \frac{B^2r^2}{2V} \tag{7.1}$$

where m = mass, z = charge, B = magnetic field, r = radius of curvature of ion trajectory, and V = electrical potential.

At constant magnetic field and accelerating voltage, each value of m/z will correspond to a radius of curvature r. Thus, after separation in the magnetic field, singly charged ions of mass m_1, m_2, m_3. . .m_n will be focused at $f_1, f_2 f_3$. . . .f_n. A series of detectors at different foci can then be used, or alternatively the magnetic field may be varied continuously, so as to bring different ions to focus at f_1 in turn. Magnetic scanning is more useful than electrical scanning, as the latter gives decreased sensitivity as accelerating voltage is decreased. The upper scan-limit is constrained by magnet technology, but scanning up to 2500 is easily

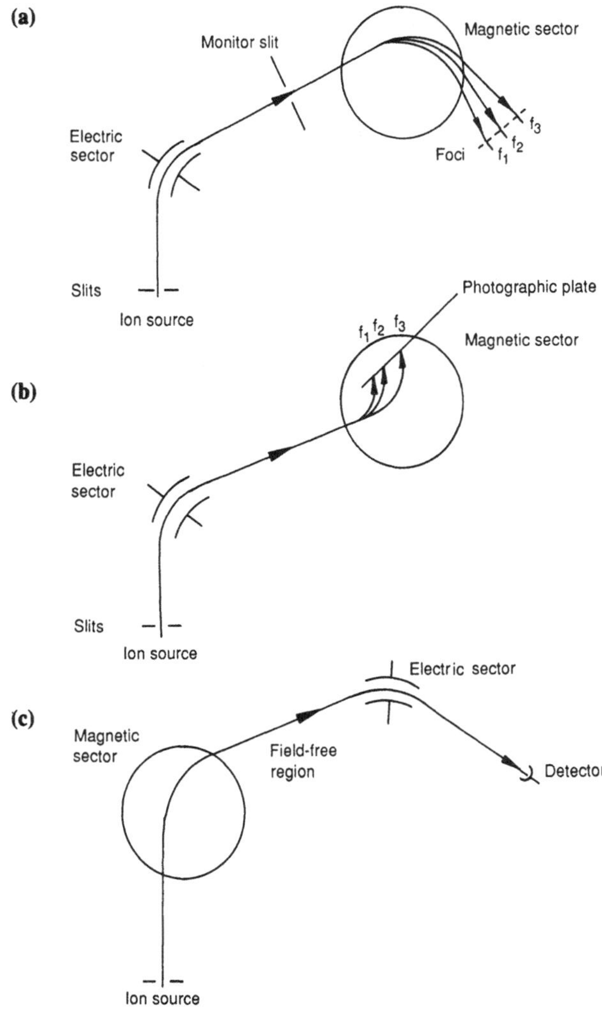

Figure 7.3 Magnetic sector mass spectrometer configurations. Ion flight paths in double-focusing instruments with (a) Nier–Johnson geometry, (b) Mattauch–Hertzog geometry, and (c) reversed geometry. The electric field is in the plane of the paper, the magnetic field is perpendicular to the page. Redrawn from Rose and Johnstone (1982).

performed. Reversed-geometry instruments (i.e. with magnetic sector preceding the electric sector) are mainly used for the detection of metastable ions.

In a single-focused instrument there is no electric sector and thus no focusing for kinetic energy. Resolution is decreased, but integer *m*/*z* values up to 2000 can be distinguished. Double-beam instruments have two adjacent sources and two detectors; this arrangement permits direct comparison of sample and reference compounds.

7.3.2 Quadrupole mass spectometers

In a quadrupole mass analyser, four precisely aligned parallel rods are electrically connected in opposite pairs (Figure 7.4); a dc voltage together with a superimposed rf potential is applied. Ions in the analyser undergo complex trajectories, and with the rf/dc voltage ratio constant, voltages are varied to effect the separation of ions. At any time only one mass can traverse the system, hence the term 'quadrupole mass filter'. Until recently, its capabilities were similar to those of a single-focusing magnetic sector mass spectrometer, with maximum resolution of about 2000 and highest detectable mass of about 1200; the mass range of newly available quadrupoles extends to 4000. It has the advantage of providing the facility for fast scanning of the whole mass range (in a few ms). Many magnetic sector machines cannot do this, because eddy currents in a magnet become problematic at scan rates greater than about 0.1 s per decade (e.g. m/z 10 to 100). The introduction of non-ferrous core electromagnets has improved this specification (Bateman, 1985). Quadrupole mass spectrometers are also cheaper, smaller and more robust than magnetic sector machines. Their rapid scanning capability makes them ideal for those combination techniques (e.g. GC/MS and GC/HPLC), where narrow elution zones from chromatographic columns are sequentially monitored.

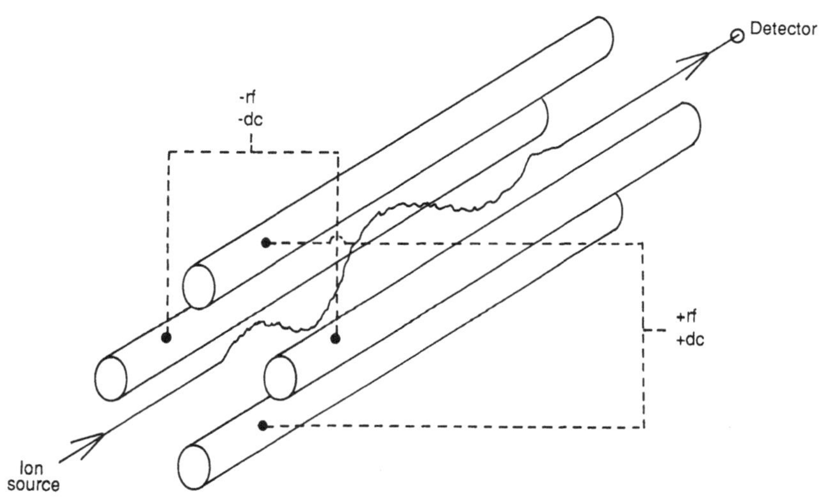

Figure 7.4 Quadrupole mass analyser. At any instant, only those ions with a single value of m/z escape collision with the rods; their complex trajectory from ion source to detector is indicated.

7.3.3 *Time-of-flight mass spectrometers*

These instruments separate ions of different masses by virtue of the different velocities acquired after acceleration through an electrical potential V (Cotter, 1989). Since $zV = \frac{1}{2} mv^2$, the velocity (v) of an ion of mass (m) is $v = (2zV/m)^{\frac{1}{2}}$. Thus the ion velocity is mass dependent, and passage through a field-free region will give different arrival times (Figure 7.5). As these times are very short, cathode ray tube detection is employed. Often used with a ^{252}Cf ion source, or with pulsed laser sources, time-of-flight mass spectrometers are used for detection of ions of very high mass. The resolving power is low, but the rapid response times are ideal for studies of fast reactions.

7.3.4 *Ion storage mass spectrometers*

Two types of ion storage techniques are becoming more widely used. Fourier transform ion cyclotron resonance mass spectrometers have high resolution and a good mass range, and use crossed or alternately-pulsed

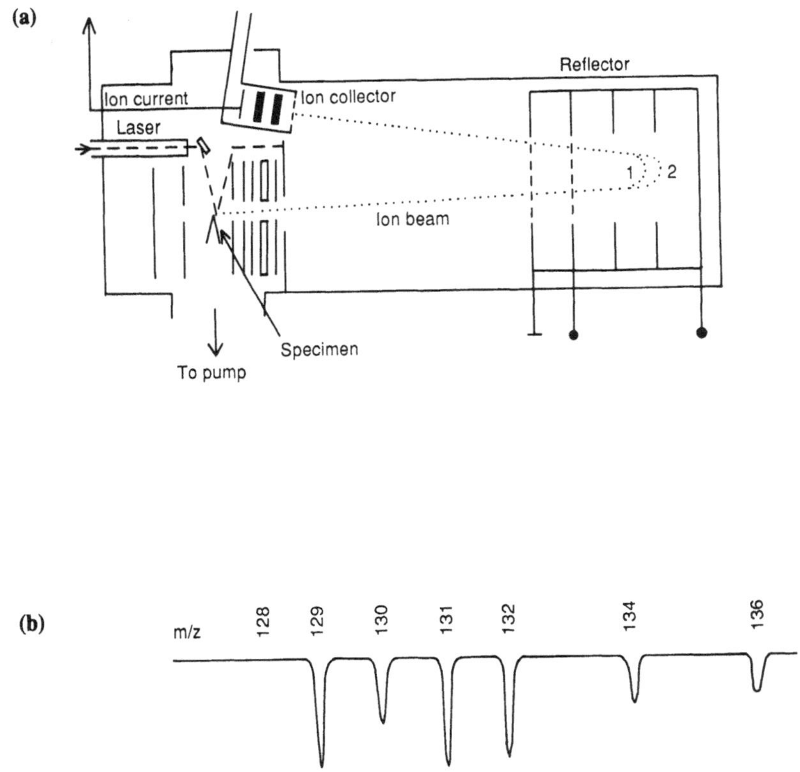

Figure 7.5 (a) Reflecting time-of-flight mass spectrometer. 1 = slow ions; 2 = fast ions. (b) Time-of-flight mass spectrum of xenon. Time base 0.05 μs cm^{-1}; xenon at resolution 420.

electric and magnetic fields (Asamoto, 1988). They are used mainly to investigate ion–molecule interactions.

The latest type of analyser in commercial production is the ion trap (Allison and Stepnowski, 1987; Figure 7.6); it is cheap and is compatible with GC. It is capable of fast response times and parts-per-billion (ppb) detection limits (e.g. for organic compounds sampled in aqueous solution via a hollow-fibre membrane probe, Lister *et al.*, 1989).

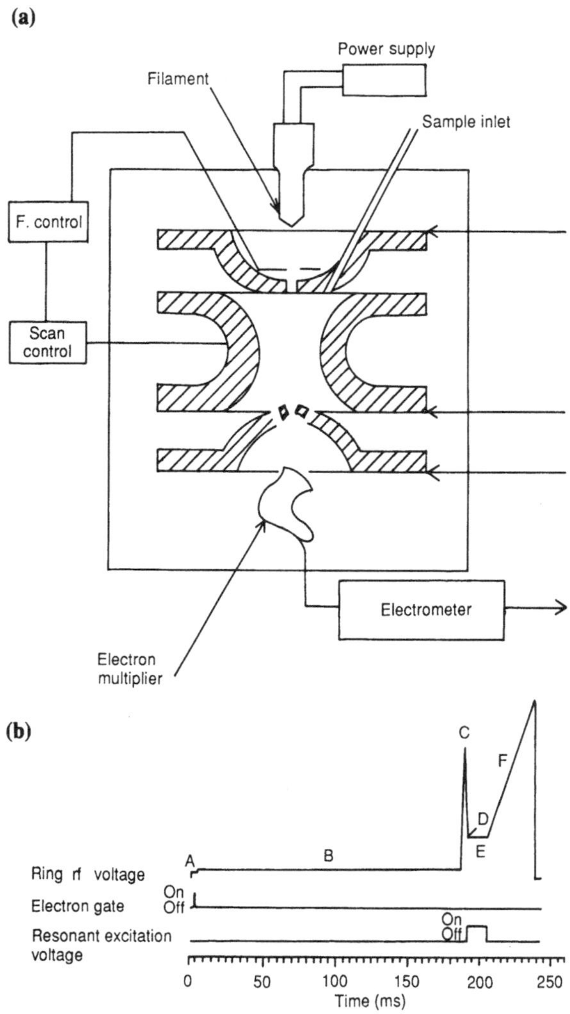

Figure 7.6 (a) Section through an ion-trap mass spectrometer. (b) Sequence of processes in the ion-trap set in daughter scan mode. From Johnson *et al.* (1990). A, Electron impact (EI) of chemical ionization (CI) reagent gas and formation of CI reagent ions. B, Reaction of CI reagent ions with neutral sample molecules to form sample ions. C, Selection of parent ion. D, Selection of daughter ions' mass range. E, Collision-induced dissociation (CID) of parent ion. F, Scan resultant daughter spectrum.

7.4 Detection and data acquisition

Detectors are of four types; the photographic plate, Faraday cup, electron multiplier or scintillation counter. The photographic plate enables simultaneous detection of ions at different focal points (e.g. in a double-focusing mass spectrometer of Mattauch–Herzog geometry). Quantitation necessitates subsequent microdensitometry scanning of the bands produced after development of the plate. Faraday cup detection involves collection of ions on a metal cup connected via a resistor to ground potential; the resulting current flow is a measure of ion abundance. This device is much less sensitive than an electron multiplier which consists of about ten electrodes (dynodes), in series. At each dynode stage, electron multiplication occurs, so that an overall gain of about 10^6 over the initial ion current can be achieved. The channel electron multiplier array is an extremely sensitive device consisting of a network of electron multipliers, and is used instead of the photographic plate. In the Daly detector, scintillations produced by collision of ions with a phosphor screen are monitored by a photomultiplier. A new focal plane charge detector with high collection efficiency, resolution and sensitivity and low noise promises 10^3 gain over currently used systems (Birkinshaw, 1992).

Signals from the detector are collected using a chart recorder, mirror galvanometer–UV sensitive paper recorder, or computer. A computer also controls continuous rapid scanning of the mass spectrometer from the time of introduction of the sample, data acquisition, storage and processing. 'Library searching' enables automated comparison of the spectrum of an unidentified compound with standard spectra stored on disc.

7.5 Tandem mass spectrometry

In tandem mass spectrometry (MS-MS), two mass spectrometers are coupled together in order to provide separation and analysis with extra selectivity and sensitivity (McLafferty, 1983). A gas collision cell between the two spectrometers provides the facility for fragmentation of any ion selected by MS-I to yield characteristic product ions to be analysed and detected by MS-II (daughter ion spectra). By varying the amount of energy injected into ions in the gas collision cell, the extent and nature of fragmentation can be varied depending on the analytical or structural information required. The type of gas used (e.g. argon, helium, xenon, nitrogen) and the gas pressure are the experimental variables employed. Several types of tandem MS machines are commercially available. In the widely used 'Triple Quadrupole' (Yost and Enke, 1979; Figure 7.7), both MS-I and MS-II are quadrupoles, whereas the 'Hybrid' instrument has a double-focusing instrument as MS-I, and MS-II is a quadrupole. The gas collision cell is, in both cases, an rf-only quadrupole.

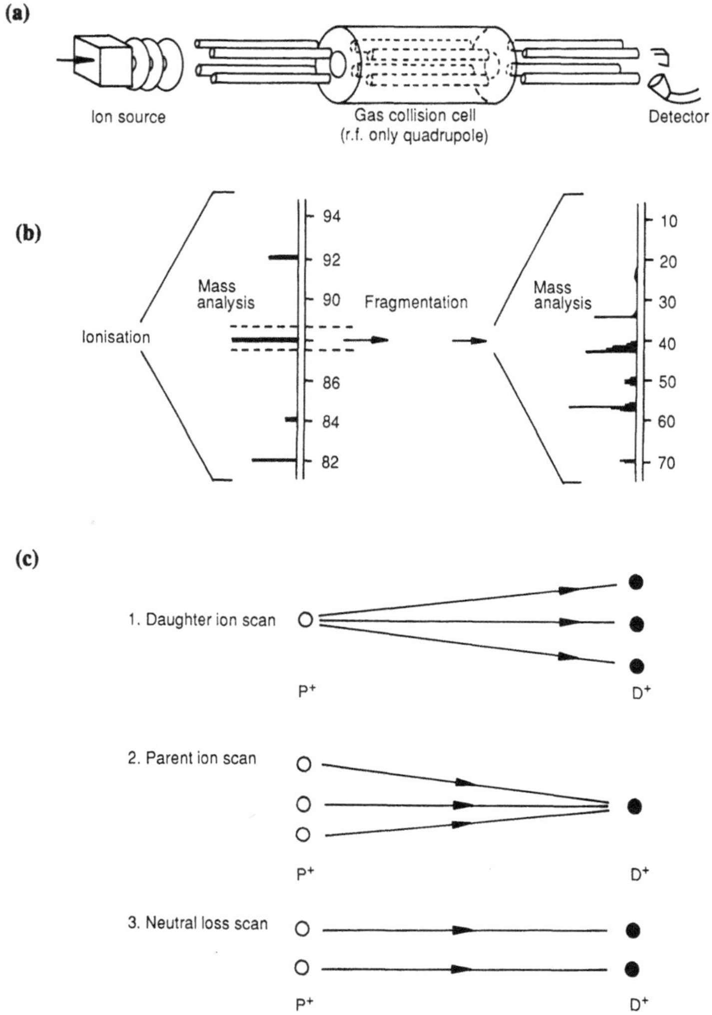

Figure 7.7 Tandem mass spectrometry: (a) triple quadrupole instrument, from Yost and Enke (1979), (b) daughter scan mode, (c) alternative scan modes.

Finally, MS-I may be a double-focusing analyser, and MS-II either a second double-focusing or a single-focusing device.

The majority of applications of tandem mass spectrometry to the determination of trace analytes in highly complex mixtures has involved triple quadrupole machines, and parts-per-trillion limits of detection have been reported for minor constituents of blood, urine or environmental samples. In a triple mass spectrometer, there are three fundamental control parameters. These are the mass selected by the first mass analyser, the presence or absence of gas in the collision cell and the mass selected by the second mass analyser. Computer controlled scanning of MS-I only, gives a normal mass spectrum. A scan of the third quadrupole while the first mass filter passes a specific mass, produces a spectrum of all the daughter ions from the selected parent ion. With collision gas absent this mode gives measurement of 'metastable' ion decomposition. The spectra of all parent ions that fragment to produce a given daughter ion is observed by scanning the first mass filter with the third quadrupole fixed on the daughter mass. Simultaneous scanning of both mass filters with a constant offset in mass between them, gives neutral loss spectra, and thus enables identification of those classes of chemical compounds which undergo characteristic mass losses in the collision cell (e.g. alcohols, 18 amu; acetylated species, 42 amu; and carboxylic acids, 44 amu).

In structure elucidation applications, a complete fragmentation map may be constructed by successive recordings of the mass spectrum of each fragment ion of a compound. For analysis of mixtures, the molecular ions for each component can be produced by soft ionization (chemical ionization or field desorption), and then separated by the first mass analyser. The molecular ion species can be selected for one component at a time. Each can then be fragmented and analysed by scanning the second analyser. Extra selectivity and discrimination eliminates 'chemical noise', and often improves the detection limit without the time delays associated with an initial chromatographic separation (see section 7.7.2).

The quadrupole ion trap mass spectrometer (ITMS) provides a new approach to instrumentation for tandem mass spectrometry (Johnson *et al.*, 1990). Whereas in the triple quadrupole or sector MS-MS instruments several sequential processes (ionization of sample molecules, mass selection of parent ions. collision-induced dissociation (CID) of parent ions with neutral gas molecules to produce daughter ions, and mass analysis and detection of these daughter ions) are conducted in separate regions of the instrument, it is now possible to achieve all this in the same physical space. This technique can be regarded as MS-MS in time (Figure 7.6).

Control of the progress of processes is achieved by controlling the rf voltage level applied to the ring electrode and the resonant excitation or (collision-induced dissociation) CID voltage applied across the endcap electrodes. For chemical ionization, the rf voltage is set to a low value so that the low m/z CI reagent ions are efficiently stored (period A). Then the ring electrode rf voltage is increased to allow storage of sample ions produced from ion–molecule reactions with CI (period B). All ions of m/z below that of the parent ion are ejected from the ion trap by raising the rf voltage on the ring electrode (period C). The trapping rf voltage is then lowered to a value appropriate for storing daughter ions of interest (period D). Tandem mass spectrometry (MS-MS) is achieved during period E by application of an ancillary ac (resonant excitation) voltage across the endcap electrodes. Ion resonance gives increased kinetic energy and collision with He buffer gas (1 mTorr) normally present. Provided that the gain of internal energy is sufficient a parent ion can undergo CID to produce daughter ions. These daughter ions will be trapped if their *m/z* lies above the low *m/z* cut-off of the ion trap. Finally a scan of the trapping rf voltage gives sequential ejection into the detector and thus mass analysis of the daughter ions leaving the trap (period F).

7.6 Ionization techniques

7.6.1 Electron impact ionization (EI)

The most common type of ion source uses a beam of electrons. Both positive and negative ions are formed by electron detachment and collision respectively; the latter are only minor products usually. Generally positive ions are then analysed. In the ion source, electrons are produced by thermionic emission from a hot-wire filament, accelerated through 70 V, and collected on an electron trap after passing through the ionization chamber. Feedback control ensures stability of filament temperature and hence electron emission. Ions emitted are focused into the analyser. Electron impact ionization produces extensive fragmentation to give cracking patterns characteristic of the molecular species and for that ionizing voltage employed.

7.6.2 'Soft' ionization techniques

In electron impact ionization, it is difficult to control the amount of energy transferred to the sample from the electron beam. In those cases where extensive fragmentation occurs (e.g. for highly branched-chain aliphatic compounds) at low energy, the abundance of the molecular ion will be extremely low. Furthermore, many polar compounds have a low vapour pressure, and cannot be volatilized without undergoing thermal decomposition. Although in many cases, preparation of a volatile derivative (e.g. esterified or silated derivative) may alleviate these problems, it is often necessary to employ a 'soft' ionization method which produces minimal fragmentation. A wide choice of soft ionization techniques now exists; these include chemical ionization (CI) and desorption chemical ionization (DCI), field desorption (FD), laser desorption (LD) and ^{252}Cf desorption. Fast atom bombardment (FAB), thermospray ionization and electrospray ionization are dealt with under sample introduction (Section 7.7.2). These ionization methods are excellent for determinations of relative molecular mass, since ions such as M^+, MH^+, $(M\text{-}H)^-$ predominate.

7.6.2.1 Chemical ionization. Chemical ionization is of great value in providing information complementary to that obtainable from EI. Rapid switching from one mode to the other is now widely available, and both positive and negative ions may be analysed. The ion source has an inlet for the reactant gas and a facility which allows operation at 10^{-5} Torr for EI or 0.5 Torr for CI. Gases used include methane, ammonia and freon (negative ions). Desorption CI (Hunt *et al.*, 1977) involves flash evaporation from a thin wire held in the reactant gas plasma and is invaluable for thermally-labile compounds.

7.6.2.2 Field desorption. Field desorption consists of the use of an intense electric field (10^5 V cm^{-1}), Beckey (1977). It is still pre-eminent for structural determinations of thermally-labile compounds, e.g. to detect single amino acid changes in proteins (Matsuo *et al.*, 1981).

7.6.2.3 Laser desorption. Laser desorption uses pulsed laser beams for ionization (Posthumus *et al.*, 1978). Matrix assisted laser desorption ionization (using a matrix of derivatives of cinnamic or nicotinic acids that strongly absorb the photon energy of the YAG (yttrium–aluminium–garnet) or N_2 laser lines, at 266 or 337 nm respectively, but do not affect the analyte) with time-of-flight analysis (LD-TOF) has been used for analysis of as little as 1 pg of singly-protonated protein (M_r < 100 000; Karas and Hillenkamp, 1988). Multi-photon ionization (MUPI-MS) uses a tunable laser for wavelength selective optical ionization of peptides, sugars, chlorophylls, and so on (Grotemeyer and Schlag, 1988). Controlled fragmentation at low laser intensity gives predominantly molecular ions; increased fragmentation can be promoted by increasing photon flux. Laser activated mass analysis (LAMMA) has been used to quantitate Na^+ and K^+ in single bacterial cells (Lindner and Seydel, 1983). Resonant ionization mass spectrometry (RIMS) using titanium–sapphire lasers and optical paramagnetic oscillators has recently been described (Ledingham and Singhal, 1991). This system presents an ultrasensitive technique of enormous potential for inorganic atomic analysis using solid state laser devices with broad tuning ranges but without the disadvantages associated with dye lasers. New IR, visible and UV radiation sources that provide means of volatilization and ionization of very large molecules will revolutionize biochemical methodology.

7.6.2.4 ^{252}Cf desorption. ^{252}Cf desorption uses a 104 MeV fission fragment to desorb intact molecular ions from solid surfaces (Torgerson *et al.*, 1974; Jungclas *et al.*, 1991). This method has been used for non-volatiles, solid samples, quasi-non-destructive analysis of rare materials, and structural determinations of peptides and proteins (Roepstorff, 1987).

In those studies where structural information is required as well as molecular masses, increased fragmentation is required, and this can be produced by soft ionization procedures used in conjunction with collision-induced dissociation (CID) techniques.

7.7 Inlet systems: sample introduction

7.7.1 Gases and votaile compounds

7.7.1.1 Capillary inlets and leak valves. Where gases or volatile liquids are in plentiful supply e.g. (< 1 mg), controlled introduction from a heated reservoir provides a simple introduction method. A small batch of the sample (e.g. 1 cm^3 gas or vapour at 1 atmosphere) is expanded into a pre-evacuated vessel (typically of 1l volume). Continuous sampling from a gas stream is used when gas composition is changing and has to be continuously monitored. Where the sample is at less than atmospheric pressure (e.g between 10^{-8} and 0.1 atmospheres), a capillary inlet suffices to drop the pressure to within the working range of the high vacuum. For higher pressures (0.1–100 atmospheres), the restricted flow requires a by-pass pump and leak arrangement. Continuous sampling requires careful calibration with a known standard mixture of composition similar to that of the sample because of possible viscosity differences between components. Similar problems of sample fractionation are also likely with other techniques, e.g. use of a leak valve. This device provides an extremely useful means of continuous sampling over a wide pressure range (10^{-8}–10 atmospheres). Another method of continuous sampling uses a jet separator. The sample gas or vapour, diluted in a carrier gas (usually helium) is drawn through a narrow jet into a reduced pressure region. An aligned second jet inlet into the mass spectrometer receives an enriched sample as the lighter helium gas diffuses away.

7.7.1.2 Membrane inlets. An extremely useful and versatile approach to the problem of pressure reduction at the inlet employs a thin (e.g. 10–100 µm) permeable membrane. An enormous range of membranes, with well-documented permeability properties, is available (Bixler and Sweeting, 1971; Henis and Tripodi, 1983), but for many purposes, investigators need not look beyond the choice of either polytetrafluoroethylene (PTFE) (permeable only to fixed gases), or silicone rubber (permeable to both gases and volatiles). For gases, transit times are rapid enough at ambient temperatures, whereas heated membrane inlets (at temperatures up to 100°C) give increased transfer rates for volatiles (Bohátka *et al.*, 1983). Membrane inlets may be used for gas samples or for liquid phase measurements of dissolved gas or volatile species (Degn, 1992). Two types of configurations have been employed (Fig 7.8); membrane inlet probes (lengths up to several metres) are useful for *in situ* monitoring of headspace gases or directly in the stirred liquid phase of a fermentor (Lloyd *et al.*, 1985; Lloyd and James, 1987; Cox, 1987), environmental sites (Lloyd *et al.*, 1986) or within the digestive tract of

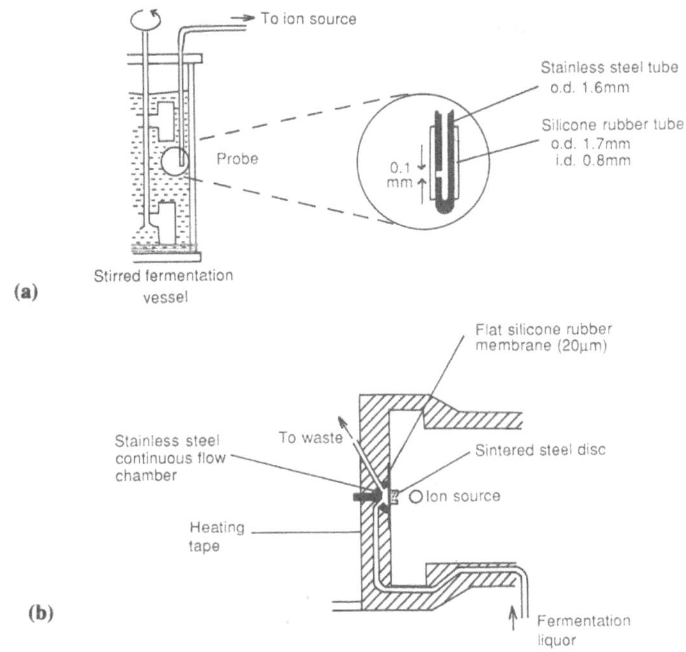

Figure 7.8 Membrane inlets. (a) Membrane inlet probe for continuous dissolved gas monitoring in a fermenter. (b) Continuous flow monitoring of volatiles in fermentation broth passing over a heated membrane inlet.

animals (Lloyd *et al.*, 1992a). A supported membrane sheet can be used as direct interface between mass spectrometer and reaction vessel (Lloyd and Scott, 1983; Degn *et al.*, 1985; Lauritsen *et al.*, 1991). Where maximum sensitivity for measurement of volatiles is required, the heated membrane interface is located as closely as possible to (or even within) the ion source (Bier and Cooks, 1987; Lauritsen, 1990; Hayward *et al.*, 1990; Bier *et al.*, 1990). Membrane inlet mass spectrometry also provides a method for the measurement of water in organic solvents (Bohátka and Degn, 1991; Degn *et al.*, 1992).

7.7.1.3 Ion mobility spectroscopy (IMS). This technique involves ionization of neutral analyte vapours by passage through a ^{63}Ni ring. Ions are electrically-gated, and drift through an anti-parallel flow of buffer gas (air or N_2). The ions are focused by an electrical field about the heated, cylindrical drift region, and are registered by a Faraday cup detector. The entire process is conducted at ambient or near ambient pressure and therefore ion formation involves atmospheric pressure ionization chemistry. Characterization of ions is by corrected drift times (typically ms) or ion mobilities.

7.7.1.4 Combined gas liquid chromatography–mass spectrometry (GC-MS). Direct coupling from a chromatographic column mass spectrometer can be used, provided that the column liquid phase is of low volatility and high thermal stability, and accounts for less than about 10% of the packed column volume. GC-MS has a long established record of applications in all sorts of situations where complex mixtures of organic compounds are to be characterized. With efficient vacuum pumps, the mass spectrometer can cope with carrier gas flow rates of about 2 ml min^{-1} or less. A quartz-lined metal capillary ensures against catalytic decomposition of sample at elevated temperature. An alternative interface splits the sample stream so that only a small proportion of the GC effluent enters the ion source: parallel detection by flame ionization is then feasible. Enrichment of sample before entry into the MS uses three main methods: (i) silicone rubber membranes can be used selectively to partition the organic sample from carrier gas (Llewellyn–Littlejohn membrane separation), (ii) selective effusion of helium through a porous (sintered glass) partition (Watson–Biemann effusion separation), or (iii) differential diffusion of solute and carrier gas after passage through a small jet into a vacuum (Rhyage jet separation). Multiple jet stages and means of deflecting He away from the ion source avoids the build-up of unacceptably high pressures and damage to the ion source. Beam deflection time-of-flight mass spectrometry can provide mass spectra at a scan file generation rate much higher than that possible with quadrupole or magnetic sector devices but at comparable detection limits (Watson *et al.*, 1990).

7.7.2 Liquids

7.7.2.1 Liquid chromatography–mass spectrometry (LC-MS). When working with on-line liquid chromatography eluent, the physical constraints placed upon the mass spectrometer system are much greater than those for GS-MS. The best available pumps cannot deal with the 1.2 l of gas per minute at 1 atmosphere produced from 1 ml min^{-1} of water. To solve this problem it is possible, (i) to match HPLC flow to suit the mass spectrometer, (ii) remove solvent before arrival at the mass spectrometer source (e.g. in a moving belt interface, Figure 7.9), (iii) use the ionizing characteristics of (some) solvent removal processes, or (iv) operate the mass spectrometer at atmospheric pressure. In direct liquid inlet systems (DLI), only a very low flow of eluent is permitted to enter the source. To achieve this, the stream is split, or use is made of a micro-HPLC system.

7.7.2.2 Fast atom bombardment (FAB). In fast atom bombardment a sample dissolved in a liquid matrix (e.g. glycerol) is exposed to a beam of neutral (8 keV) particles (e.g. argon or xenon atoms). Solvent removal is

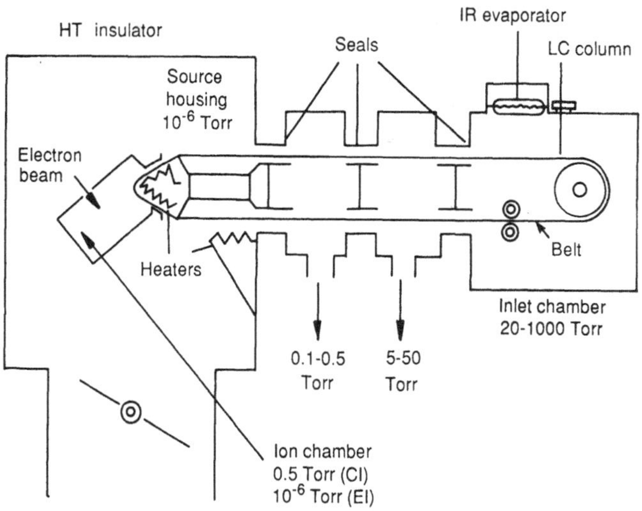

Figure 7.9 Moving belt interface. From D. S. Millington, unpublished.

often achieved by use of a moving-belt interface. In this device spray deposition of the eluent on a Kapton (polyimide) moving belt (Figure 7.9) precedes flash heating within the ion-source housing and enables freedom to operate the source in any ionization mode (EI, CI or FAB). Polar and charged molecules give the most intense spectra; amines, salts of acids, amides and alcohols are good examples, and the method has been used for protein sequencing (Biemann, 1986). Continuous flow analysis of aqueous samples has been employed for peptides (M_r 1500, 5 pmol; Caprioli and Fan, 1986). Eluent flow is mixed with glycerol and allowed to form a film at the end of a probe in a FAB source (Caprioli, 1990).

7.7.2.3 Supercritical fluid chromatography–mass spectrometry (SCF-MS). The principle of this technique is to use a supercritical fluid (e.g. CO_2), with or without a proportion of modifying solvent (e.g. methanol or acetonitrile) as the solvent, and either a HPLC packed column or bonded-phase capillary GC column in a temperature controlled oven. Variation in the polarity of the solvent is achieved either by changes in pressure or by changing the proportion of modifying solvent. Coupling to a mass spectrometer provides a powerful method, as SCF separations are fast and give narrow peaks, and low gas burdens. A heated capillary-belt interface has been used to give excellent separation of non-polar non-volatiles (D. E. Games, personal commun.).

7.7.2.4 Capillary zone electrophoresis–mass spectrometry. The new techniques of high performance capillary zone electrophoresis (HPCE) enable highly efficient separations of biomolecules from minute sample volumes containing complex mixtures.

These capillary columns are easily coupled to a mass spectrometer. The correspondence between HPCE and electrospray ionization flow rates and the fact that both techniques are primarily used for ionic species makes for an extremely sensitive combination, e.g. for tryptic digests (Lee *et al.*, 1989).

7.7.2.5 Thermospray. Ionization as a consequence of solvent removal underlies the technique of thermospray (TSP)-LC interfacing. The solvent contains an ionic buffer (usually ammonium acetate), and this passes down an electrically-heated stainless steel capillary tube. For ionic substances this buffer is not necessary, and non-polar materials may be ionized by combining a TSP interface with a filament and/or plasma discharge source. Thermospray enables specific analyses of non-volatile and polar compounds without derivatization, e.g. 50–100 pg indolic compounds (Auriola *et al.*, 1990).

7.7.2.6 Electrospray (ESI). In this technique (Figure 7.10), injection of the LC effluent into a flow of nitrogen through a metal capillary held at a potential of several kilovolts relative to the surrounding chamber walls results in a spray of 1 μm drops. Desolvation of the fine spray in a stream of warm gas leads to the ejection of ions at temperatures and pressures close to ambient. Electrospray enables rapid (5 min) characterization of biological molecules directly from liquids. For example, quadrupole analysis of

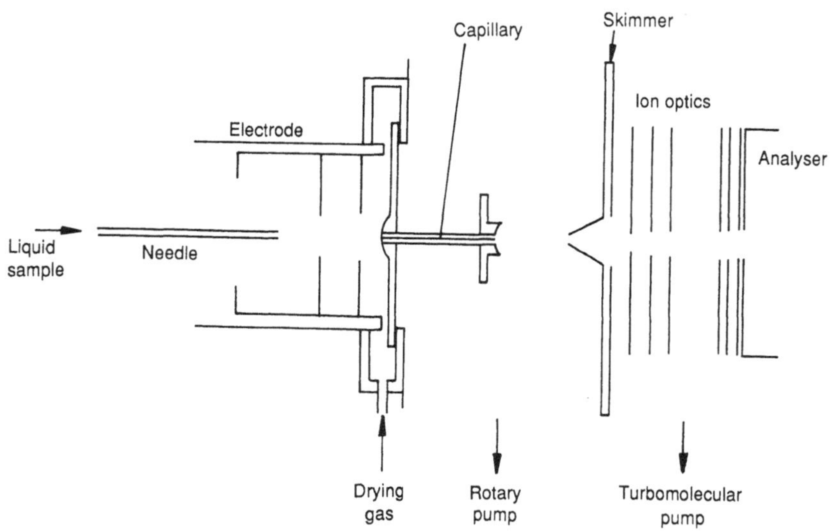

Figure 7.10 Electrospray ionization.

multiply-protonated protein fragments can give molecular mass determinations to better than 0.01% accuracy (Fenn *et al.*, 1989) for proteins $M_r <$ 130 000 at low picomole to femtomole levels. Sequence analysis methods for peptides by MS-MS analysis have been described (Smith *et al.*, 1990; Loo *et al.*, 1990). Characterization of post-translationally modified proteins is becoming an extremely important application of this technique (Biemann and Scoble, 1987; Poulter *et al.*, 1990).

Electrospray ionization mass spectra of oligonucleotides (up to 61-mer) and tRNA utilizes negative ion analysis; M_r values up to 10 000 can be accurately determined (Loo *et al.*, 1989). Electrospray is easily coupled to HPLC or HPCE at low rates of 1–40 μl min^{-1}.

7.7.3 Solids

A solid sample insertion probe enables introduction of solid samples directly into the vacuum system: studies of surface chemistry generally lie outside the scope of this review. However, one method that uses such a probe, pyrolysis mass spectrometry, has widespread applications in biology.

7.7.3.1 Pyrolysis mass spectrometry (Py-MS). Pyrolysis mass spectrometry provides a rapid analytical fingerprinting technique for non-volatiles. A few μl of suspended material (e.g. organisms) is dried and then heated (e.g. to 770°C for 0.6 s). The controlled thermal decomposition of matter in an inert environment yields volatile fragments characteristic of the original material. Multivariate pattern recognition is then used to compare and contrast spectra. Automated sample processing gives complete analysis time of two minutes. This technique has been employed extensively for identification of microorganisms (Gutteridge and Puckey, 1982), xenobiotics (Toth and Snyder, 1991), and quality control in food science (Aries and Gutteridge, 1987).

7.8 Applications in food science technology

This section details just a few examples of mass spectrometry applications of potential usefulness in the food process industry (Gilbert, 1987).

7.8.1 Membrane inlet quadrupole mass spectrometry

Gas analysis is widely used, and a number of specialized devices are employed. Mass spectrometry offers many advantages, and small inexpensive membrane inlet quadrupole instruments provide versatile, sensitive and specific means for continuous monitoring of multiple species, in either gas or liquid phases. Thus simultaneous measurements of O_2 and CO_2 (limits ~ 0.5 μM), provide details of respiratory states of microbial cultures. H_2, CO_2 and CH_4 can be simultaneously measured in anaerobic fermentations; denitrification (N_2O and N_2), and sulphate reduction (H_2S) can also be studied (Lloyd and Scott, 1983; Lloyd *et al.*, 1985, 1992a). These methods may easily be applied to monitor possible changes in gas composition in inert atmosphere food storage systems. The mass spectrum of air is shown in Figure 7.11.

Water activity has been sensitively measured at m/z = 18 using a polyethylene terephthalate (Melinex 800, 12 μm thick; ICI, Welwyn Garden City) membrane inlet probe (Figure 7.12; Bohátka and Degn, 1991). This system can even be used to measure water in organic liquids, as the membrane is impermeable to hydrophobic solvents (Degn *et al.*, 1992). In some applications combined measurements of gases and water might be useful (e.g. using a silicone rubber membrane interface).

Low molecular mass volatiles can also be monitored through a silicon rubber membrane using a simple membrane inlet quadrupole mass spectrometer. Figure 7.13 shows the continuous monitoring of CO_2 and ethanol production in a culture of yeast. Anaerobic conditions give increased glycolytic rates due to the Pasteur effect control systems (Lloyd and James, 1987). Many studies on yeast fermentation have been reported (Pungor *et al.*, 1980; Weaver *et al.*, 1980). Heinzle *et al.* (1983) performed continuous and simultaneous on-line analyses of fermentation products in gas (capillary inlet) and liquid phases (125 μm silicone membrane) in a 7-litre continuous culture of yeast. Transients occurring when process conditions were altered (gas flow rate or dilution rate) could be faithfully recorded, and oscillations in dissolved gases and in ethanol concentrations were observed. The ethanol signal required daily calibration by reference to GC, and difficulties were experienced in the analysis of acetaldehyde due to its low concentration and the similarities in fragmentation patterns of

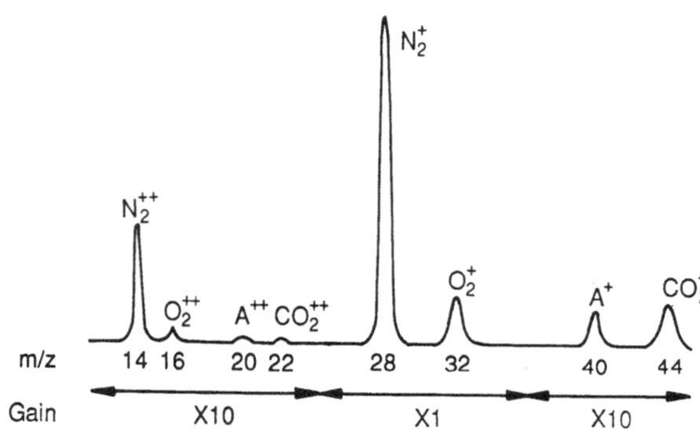

Figure 7.11 Mass spectrum of air.

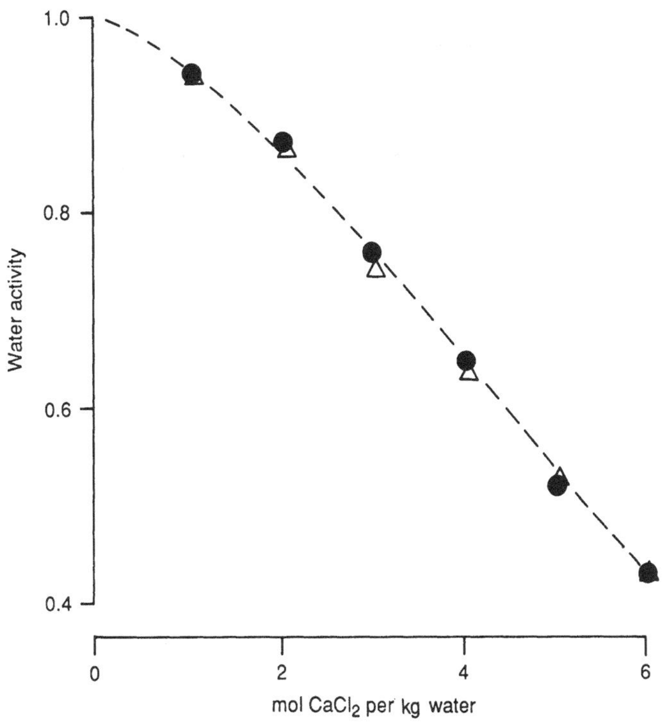

Figure 7.12 Mass spectrometric measurement of water activity through a polythene terephthalate membrane. From S. Bohátka and H. Degn, unpublished experiment. ●, calculated values; △, mass spectrometric determination.

the two compounds. Scale-up to production scale fermenters (100 m^3) used for the production of yeast biomass from molasses in an aerated feed-batch system has been described by Cox (1987). Reasonable agreement between ethanol measurements at *m*/*z* 31 and off-line chemical determinations was obtained, and the response time ($t_{90\%}$ = 3 min for a 60 cm long 0.6 cm internal diameter (id) stainless steel probe at ambient temperatures) was adequate for this process. Measurements of methanol, ethanol, propan-2-ol, propan-1-ol, butan-1-ol and acetone standards suggested that hydrophobicity as well as volatility is an important factor determining rate of passage through silicone membrane.

The Pasteur effect (the inhibition of glucose utilization by O_2) is important in many industrial fermentations, and a study of glycolytic mutants of *Saccharomyces cerevisiae* (Lloyd *et al.*, 1992b) provides a good example of how laboratory scale studies can obtain basic information by mass spectrometry. Determinations of ethanol at *m*/*z* 31 were more sensitive and specific than at other *m*/*z* values (29, 45, 46); more than 100 mM ethanol accumulates during the initial glucose-repressed phase of aerobic growth with medium initially containing 55 mM glucose,

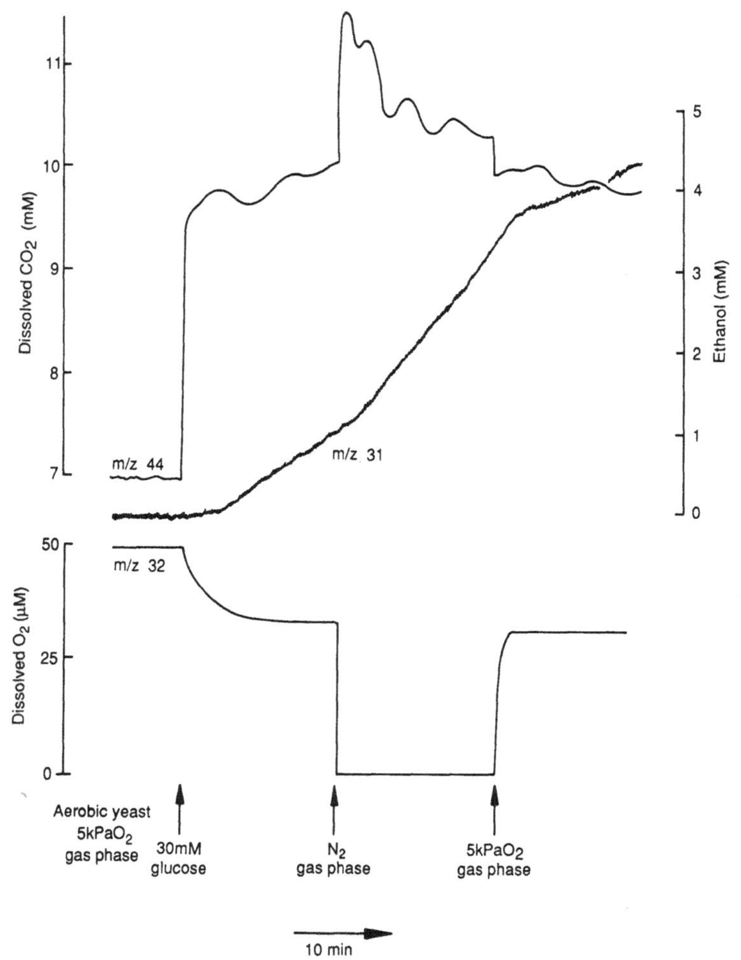

Figure 7.13 Continuous monitoring of O_2, CO_2 and ethanol (at *m/z* = 32, 44 and 31, respectively) in *Saccharomyces cerevisiae* utilizing glucose. At times indicated the suspension was made anaerobic by altering the composition of the mobile gas phase from air to N_2. From Lloyd and James (1987).

and in many strains other products (acetaldehyde, citrate, glycerol, succinate and malate measured by HPLC) are, by comparison with ethanol, only minor ones. Acetic acid can accumulate under some conditions in some yeast cultures (e.g. in those of *Brettanomyces* spp.) and, at the low pH values characteristically employed for yeast fermentations, also penetrates the silicone membrane to provide a contribution at *m/z* 31. Ethanol oxidation under aerobic conditions is rapid in most yeast strains examined, and this must be taken into account when calculating rates of production. Where rapid response times for ethanol are required, continuous flow through an external loop over a heated flat silicone rubber (Esco, thickness 100 μm) membrane gives excellent results (e.g. $t_{90\%} = 20$ s at 80°C; D. Lloyd, unpublished results). Acetone can be directly measured in milk by immersion of a silicon rubber membrane probe; each reading takes about 15 s at room temperature (D. Lloyd, unpublished data). For higher molecular mass volatiles, especially in complex fermentations where product mixtures are changing, inversion of the matrix equations necessary for the resolution of the contributions of individual molecular species becomes more error prone. Thus Doerner *et al.* (1982) were able to perform on-line monitoring of ethanol, acetone, *n*-butanol, acetic acid and butyric acid in cultures of *Clostridium acetobutylicum*. This study emphasized some of the pitfalls of a simplistic approach using EI mass spectrometry, where overlapping cracking patterns of different components make resolution difficult, and where highly mixed solvent concentrations may give cross-interacting membrane transport coefficients.

7.8.2 *Tandem mass spectrometry*

Tandem mass spectrometry provides a powerful solution to the problems of resolution of complex mixtures. The yeast *Brettanomyces anomalous* is highly resistant to food preservatives and sometimes contaminates fermented beverages. It produces strong spicy, smoke-like, clove-like, woody or phenolic odours leading to spoilage of wine, beer and cider. When unshaken cultures were maintained in the dark for 21 days in the presence of either coumaric acid, or ferulic acid, they accumulate volatile products (Heresztyn, 1986); identification of 4-ethylphenol and 4-ethyl-2-methoxyphenol was by GC-MS after a lengthy extraction and concentration procedure. The power of membrane inlet tandem mass spectrometry for rapid characterization of such compounds directly in the growth medium has recently been demonstrated (Lauritsen *et al.*, 1991). Thus incubation of organisms with coumaric acid gave a peak in the CI mass spectrum (reaction gas methane) at *m/z* 123 (protonated 4-ethylphenol) with daughter ion peaks at *m/z* 105 $[M+H–H_2O]^+$, *m/z* 95 $[M+H–C_2H_4]^+$ and *m/z* 77 $[M+H–C_2H_4–H_2O]^+$. From *m/z* 153 (protonated 4-ethyl-2-methoxyphenol) in cultures incubated with ferulic acid, daughter ion peaks were obtained at *m/z* 138 $[M+H–CH_3]^+$, *m/z* 121 $[M+H–CH_3OH]^+$ and *m/z* 93 $[M+H–CH_3OH–C_2H_4]^+$. CI mass spectra of all samples other than controls (i.e. media from yeast cultures without addition of precursors) gave an intense peak at *m/z* 89 (protonated ethyl acetate); major daughter ions at *m/z* 61, 43 and 29 correspond respectively to $[M+H–C_2H_4]^+$, $[M+H–C_2H_5OH]^+$ and $[M+H–CH_3COOH]^+$. Figure 7.14 shows the daughter ion mass spectra of components produced from coumaric and ferulic acids by *B.anomalous*; both 4-ethylphenol and 4-ethyl-2-

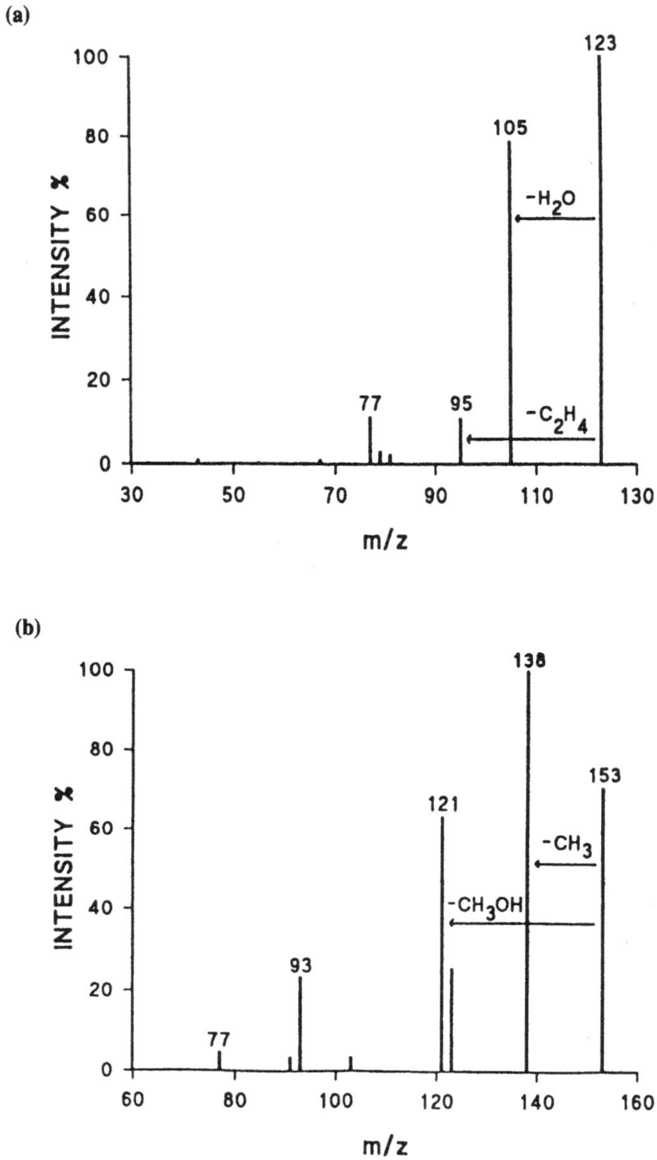

Figure 7.14 Daughter ion mass spectra of (a) *m/z* 123 protonated 4-ethylphenol from fermentation of coumaric acid by *Brettanomyces anomalous*, and (b) *m/z* = 153 from the culture containing ferulic acid. From Lauritsen *et al.* (1991).

methoxyphenol penetrate the silicone membrane interface, and can thus be monitored directly and continuously in the fermenter (Lauritsen *et al.*, 1991). The power of tandem mass spectrometry in fermentation monitoring is also well illustrated by its application to studies of butanediol and acetoin accumulation (Figure 7.15; Kotiaho *et al.*, 1991). *Klebsiella oxytoca*

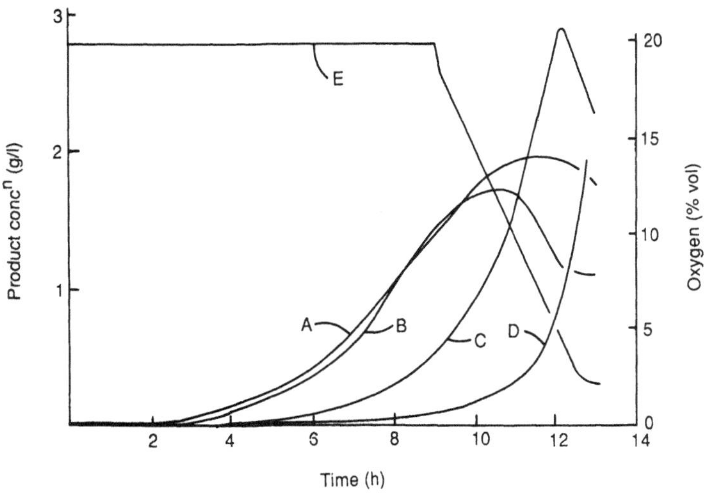

Figure 7.15 Concentrations of products and feed oxygen in a *Klebsiella oxytoca* fermentation. A, ethanol; B, acetic acid; C, 2, 3-butanediol; D, acetoin; E, O_2 (From Kotiaho *et al.* (1991).

culture supernatants were analysed through a dimethyl vinyl silicone capillary tube located within 1 mm of the electron beam of the ion source of a Finnigan MAT TSQ 4500 triple quadrupole mass spectrometer (Bier and Cooks, 1987). Daughter spectra from *m/z* 91 and 89 ions in CI mass spectra (methane) confirmed the identities of 2,3-butanediol and 3-hydroxy-2-butanone (acetoin) in the fermentation broth; ethanol was also present. High sensitivity and fast response times were achieved by close proximity of the heated membrane to the ionizing region. Fall times (defined as the time taken after flow-through at 0.5 ml min^{-1} for the signal to fall from 90% to 10% of the maximum) were measured for 2-pentanone at high temperatures (e.g. at 190°C which gave a value of 31.7s). On-line monitoring of *K.oxytoca* and *Bacillus polymyxa* fermentations was reported by Hayward *et al.* (1990). A continuous flow probe constructed with a flat dimethylvinylsilicone membrane, and with controlled electrical heating, enabled flow injection analysis of fermentation broth samples interspersed with standards. Tandem mass spectrometry confirmed that measurements of acetic acid, acetoin, 2,3-butanediol and ethanol (*m/z* 61, 89, 91 and 47 after isobutane chemical ionization) were possible without cross-interference. Collision activation spectra (MS-MS spectra) under multiple ion conditions (with Ar target gas) were used to confirm product identities and to identify unassigned peaks (e.g. those from acetone and formic acid) in the CI spectra. Further work on optimization of the performance of the continuous flow probe especially with respect to temperature and flow rates makes an important contribution to bioreactor monitoring techniques (Bier *et al.*, 1990; Katiaho *et al.*, 1991). The use of a microporous polypropylene membrane interface instead of silicone rubber in the continuous flow probe (Lauritsen *et al.*, 1992a,b) for the direct identification of volatile organic compounds in either organic or aqueous solutions promises an extension to the range of fermentation products that can be monitored on-line (Lauritsen and Lloyd, 1993).

Tandem mass spectrometry has been used extensively to measure pesticides, antihelmintics, drug residues (Finlay *et al.*, 1986) and anabolic steroids in food products: an example of its use for this type of investigation is provided by Ramsey *et al.* (1989) who were able to detect 0.5 mg kg^{-1} chloramphenicol in fish and milk.

7.8.3 GC-MS

Examples of the use of GC-MS are too numerous to be exhaustively cited here, and we mention just two types of application.

Aroma and flavour compounds important in the food industry are routinely measured by this method; many microbial (especially fungal) flavours are of commercial interest (e.g. Schindler and Schmid, 1982; Berger *et al.*, 1988). For example, after screening more than 30 basidiomycetes, a strain of *Ischnoderma benzoinum* was selected as producing by far the largest amounts of volatile metabolites; GC-MS enabled identification of 29 compounds with different sensorial properties (Berger *et al.*, 1987). Together with rapidly developing immobilization and genetic engineering techniques, new methods for food flavour production are emerging.

GC-MS has also proved an invaluable aid in investigations of process spoilage. Spoilage of food products associated with off-odours has been investigated by headspace gas analysis (e.g. vacuum-packed beef spoilage by *Clostridium* sp., Edwards and Dainty, 1987); Table 7.1 shows a typical example (Dainty *et al.*, 1989). A comparison of volatiles produced by 31 strains of pseudomonads isolated from meat stored in air at chill temperatures with those produced by reference strains of *Pseudomonus fragi* and *Ps.fluorescens* biotype I (Edwards *et al.*, 1987) amply illustrates the great discriminatory potential of MS-GC for identification purposes as well as the use of this technology in food science (Dainty and Mackey, 1992). The production of undesirable fermentation products (e.g. diacetyl) is occasionally a problem in the brewing and cider making industries (Williams and Tucknott, 1978). Table 7.2 shows typical GC-MS analysis of headspace volatiles in normal and spoiled cider (Jarvis, 1993). Other uses of GC-MS include the quantitative analysis of diethylstilbestrol in animal tissues (Covey *et al.*, 1985).

Table 7.1 Volatile compounds detected in the headspace gases of naturally contaminated, pure culture-inoculated and uninoculated samples of normal (N) and high (H) pH beef stored aerobically at + 5°C (Dainty *et al.*, 1989)

	Uninoculated	Naturally contaminated	*Hafnia alvei* AG33	*Enterobacter agglomerans* WH1	*Serratia liquefaciens* SN96	*Alteromonas putrefaciens* GB40	*Aeromonas hydrophila* SN103	*Pseudomonas fragi* NCIB 10888	Brochothrix thermosphacta C420
pH of meat	N H	N H	N H	N H	N H	N H	N H	N H	N H
Ethanol			2	2 2					
n-Propanol				2					
2-Methylpropanol	+ +	+ +	2 2	+ +	+ +	+ +	+ +	+ +	+ 2
2-Methylbutanol		1 2	2 2	2 2	2 2		2		2 2
3-Methylbutanol	+ +	1 2	2 2	2 2	2 2	+ +	+ 2	+ +	2 2
2,3-Butanediol		1	2	2	2				2
Isovaleraldehyde	+ +	1 +	+ +	+ +	+ +	+ +	+ +	+ +	+ 2
Ethyl esters		2 2						2 2	
Isopropyl esters		2 2							
Isobutyl esters		2 2				no growth	no growth		
Isoamyl esters		2 2							
1,4-Heptadiene		2 2							
1-Undecene		2 2						2 2	
1,4-Undecadiene		2 2						2 2	
Diacetyl	+ +	2	2 +	2 +	2 +	+ +	+ +	++	2 2
Acetoin	+ +	1* 2*	2 +	2 +	2 +	+ +	+ +	2* 2*	2 2
Hydrogen sulphide			2 2	2 2	2 2	2	2		
Methanethiol		2 2	2 2		2 2	2	2	2 2	
Dimethyl sulphide		2 2						2 2	
Dimethyl disulphide		2 2	2 2		2 2	2	2	2 2	
Dimethyl trisulphide		2 2	2 2		2 2	2	2	2 2	
Methylthioacetate		2 2	2 2		2 2	2	2	2 2	
Methylthiopropanoate							2	2	
Bis(methylthio)methane			2		2	2			

Sterile slices of *M. semitendinosus* were stored for 6–8 days either uninoculated or after inoculation with single bacterial strains to an initial cell density of about $\log_{10}$ 4 cm^{-2}. End cuts from the unsterilized muscles were stored as naturally contaminated samples. The pH of the normal muscles used in both experiments was 5.5–5.8; that of the high pH muscles in experiments 1 and 2; 6.3–6.5 and 6.6–6.8, respectively.

+, present at same level as in fresh, unstored muscle; 2, compound produced during storage in both experiments; 1, compound produced in one experiment only: *, compound present in one or both experiments at lower levels than in corresponding fresh sample; blank, compound absent.

Table 7.2 Volatile compounds in a normal and a diacetyl-spoiled cider (Jarvis, 1993)

Compound	Normalized peak area*	
	Normal	Spoiled
Ethyl acetate	86.1	89.2
Diacetyl	0.00	3.42
Ethyl-2-methylbutyrate	10.3	12.9
2-Methyl propanol	35.4	97.3
Isoamyl alcohol	305	213
2- and 3-Methylbutan-1-ol	503	456
Ethyl hexanoate	233	179
Hexyl acetate	1.95	6.25
Octanal	0.67	1.19
Ethyl lactate	45.9	37.9
Hexan-1-ol	35.7	29.5
Nonanol	0.79	0.86
Unknown ester/acetal (rel.mol.mass 172)	25.6	77.7
Ethyl octanoate	280	226
Heptan-1-ol	1.26	0.75
Ethyl octanoate	3.39	11.5
Decan-2-one	3.58	0.90
Benzaldehyde	1.42	1.41
Ethyl-2-hydroxy-4-methyl pentanoate	4.19	7.03
Ethyl decanoate	57.6	53.4
Decenal	7.52	0.36
Ethyl benzoate	4.70	5.66
Diethyl succinate	5.23	2.88
Unknown ester	3.19	4.93
Methionol	1.29	0.55
Undecanal	2.16	1.23
2-Phenylethyl acetate	11.4	6.42
Hexanoic acid	12.6	12.6
Ethyl dodecanoate	2.10	1.04
2-Phenylethanol	69.0	61.6
Heptanoic acid	0.00	3.39
d-Decalactone	4.39	3.39
Ethyl guaiacol	2.47	4.53
Octanoic acid	34.6	29.4
Nonanoic acid	1.01	1.21
4-Ethylphenol + eugenol	1.86	1.52

*Based on GC-MS analysis of headspace volatiles

7.9 Current and future developments

Mass spectrometry has utilized and integrated diverse methodological procedures from many fields to yield some of the most important analytical instruments now available. Capabilities for high mass analysis have been revolutionized by the introduction of (i) direct estimation by laser desorption with time-of-flight analysis (Karas and Hillenkamp, 1988), and (ii) analysis of fragments produced by electrospray ionization (Fenn *et al.*, 1989). Figure 7.16 shows the computer deconvolution of the multiply-

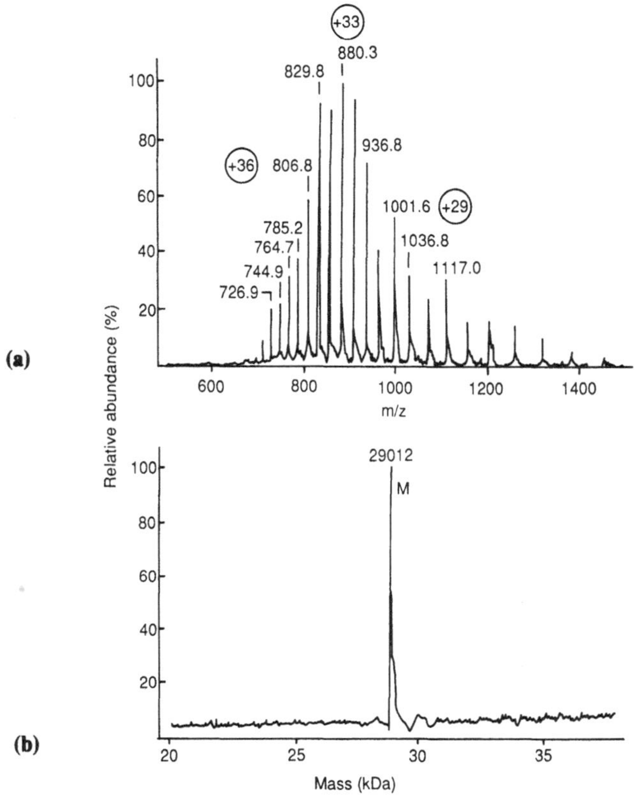

Figure 7.16 (a) Electrospray mass spectrum obtained from carbonic anhydrase (mol. wt. approx. 29 000 Da) infused into the electrospray ionization-quad system at a concentration of 3×10^{-6}M showing the multiply-protonated ion profile (not all ions have been labelled). (b) Computer deconvolution of the relative abundance versus mass/charge profile data of (a) into a relative abundance versus mass plot from which the protein mass can be read directly. From Jardine (1990).

protonated ion profile of carbonic anhydrase to give direct reading of the protein mass with an accuracy of about 0.01% (Jardine, 1990). A wide range of applications to studies of natural products has been reviewed (Burlingame *et al.*, 1988). Even so, there is evidently still a pressing need for continuing interdisciplinary collaboration, because for the most part, the outstanding capabilities of recently introduced techniques are as yet not widely appreciated, and are hence underutilized, in terms of applications. The exacting requirements of biological research also drive new developments in instrumentation. Thus a recent major shift of interest to the mass spectrometry of macromolecules has come about largely as a result of the introduction of the electrospray ionization technique which makes possible the direct 'trampolining' of biological molecules from aqueous solution as intact stable molecular ionic species. When combined with methods of unequalled analytical sensitivity at high molecular mass

Table 7.3 Sensitivities available for high moleculer mass analyses of proteins (Roepstorff, 1992).

	Sensitivity	Mass range (KD)	Mass accuracy	Mass analysis
Plasma desorption	1–10 pmol	20	0.1–0.2%	TOF
Fast atomic bombardment	1–50 pmol	10	0.05–0.2%	Sect/Quod
Electrospray	0.1–5 pmol	>100	0.01–0.02%	Sect/Quod
Matrix-assisted laser desorption	0.001–1 pmol	>250	0.01–0.1%	TOF

Figures depend on purity and concentration, and they vary from one sample to another: all are relatively soft ionization techniques.

(Table 7.3), these procedures are already able to provide more accurate and specific structural information than that available from most existing biochemical and immunological methods (e.g. electrophoretic band separations on gels cannot usually provide better than 5% accuracy for molecular mass determinations). The need for on-line analyses of highly complex mixtures (e.g. as typically the case for industrial fermentation processes) imposes new criteria for instruments in biotechnology monitoring and control. The complexity of mass spectrometric data output (e.g. from pyrolysis mass spectrometry) necessitates the implementation of new mathematical methods of analysis. Thus progress on a wide front holds promise for the continuing increased usefulness of mass spectrometric analytical procedures.

References

Allison, J. and Stepnowski, R. M. (1987) The hows and whys of ion trapping. *Analyt. Chem.*, **59**, 1072A–1088A.

Aries, R. E. and Gutteridge, C. S. (1987) Applications of pyrolysis mass spectrometry to food science. In *Applications of Mass Spectrometry in Food Science* (ed. Gilbert, J.). Elsevier, London.

Asatomo, B. (1988) Fourier transform-mass spectrometry for industrial problem solving. *Spectroscopy Internat.*, **1**, 28–37.

Auriola, S., Naaranlahti, T., Kostiainen, R. and Lapinjoki, S. P. (1990) Determination of *Catharanthus* alkaloids with thermospray liquid chromatography mass spectrometry. *Biomed. Environ. Mass Spectrom.*, **19**, 609–612.

Bateman, R. H. (1985) A novel general purpose double focussing mass spectrometer. In *Advances in Mass Spectrometry in 1985* (ed. Todd, J. F. J.). John Wiley, Chichester, p. 863.

Beckey, H. D. (1977) *Principles of Field Ionization and Field Desorption Mass Spectrometry*, Pergamon Press, Oxford.

Berger, R. G., Neuhäuser, K. and Drawert, F. (1987) Biotechnological production of flavor compounds: III High productivity fermentation of volatile flavors using a strain of *Ischnoderma benzoinum. Biotechnol. Bioeng.*, **30**, 987–990.

Berger, R. G., Drawert, F. and Hädrich, S. (1988) Microbial sources of flavor compounds. In *Bioflavour '87* (ed. Schreier, P.). Walter de Gruyter, Berlin, pp. 416–433.

Beynon, J. H. (1981) *Introduction to Mass Spectrometry*. University of Wales Press, Cardiff.
Biemann, K. (1986) Mass spectrometric methods for protein sequencing. *Analyt. Chem.*, **58**, 1288A–1215A.
Biemann, K. and Scoble, H. A. (1987) Characterisation by tandem mass spectrometry of structural modifications in proteins. *Science*, **237**, 992–995.
Bier, M. E. and Cooks, R. G. (1987) Membrane interface for selective introduction of volatile compounds directly into the ionization chamber of a mass spectrometer. *Analyt. Chem.*, **59**, 597–601.
Bier, M. E., Kotiaho, T. and Cooks, R. G. (1990) Direct insertion membrane probe for selective introduction of organic compounds into a mass spectrometer. *Analyt. Chim. Acta*, **231**, 175–190.
Birkinshaw, K. (1992) Focal plane charge detector for use in mass spectrometry. *Analyst*, **117**, 1099–1104.
Bixler, H. J. and Sweeting, O. J. (1971) Barrier properties of polymer films. In *The Science and Technology of Polymer Films* (ed. Sweeting, O. J.). John Wiley, New York, pp. 1–130.
Bohátka, S. and Degn, H. (1991) Continuous measurement of water in organic solution by membrane inlet mass spectrometry. *Rapid Commun. Mass Spectrom.*, **5**, 433–436.
Bohátka, S., Langer, G., Szilágyi, J. and Berecz, I. (1983) Gas concentration determination in fermentors with quadrupole mass spectrometer. *Internat J. Mass Spectrom. Ion Phys.*, **48**, 277–280.
Burlingame, A. L., Maltby, D., Russell, D. H. and Holland, P. T. (1988) Mass spectrometry. *Analyt. Chem. Fund. Rev.*, **60**, 294R–342R.
Caprioli, R. M. (1990) Continuous-flow fast atom bombardment mass spectrometry. *Analyt. Chem.*, **62**, 477A–485A.
Caprioli, R. M. and Fan, T. (1986) High sensitivity mass spectrometric determinations of peptides: direct analysis of aqueous solutions. *Biochem. Biophys. Res. Commun.*, **141**, 1058–1065.
Chace, D. H. and Abramson, F. P. (1989) Selective detection of ^{13}C, ^{15}N and D labelled metabolites by capillary gas chromatography-chemical reaction interface/mass spectrometry. *Analyt. Chem.*, **61**, 2724–2730.
Cotter, R. J. (1989) Time-of-flight mass spectrometry: an increasing role in the life sciences. *Biomed. Environ. Mass Spectrom.*, **18**, 513–532.
Covey, T., Maylin, G. and Henion, J. (1985) Quantitative secondary ion monitoring gas chromatography/mass spectrometry of diethylstilbestrol in bovine liver. *Biomed. Mass Spectrom.*, **12**, 274–287.
Cox, R. P. (1987). Membrane inlets for on-line liquid-phase mass spectrometric measurements in bioreactors. In *Mass Spectrometry in Biotechnological Process Control* (eds. Heinzle, E. and Reuss, M.) Plenum Press, New York, 63–74.
Dainty, R. H. and Mackey, B. M. (1992) The relationship between the phenotypic properties of food spoilage organisms and the nature of meat spoilage processes., *J. Appl. Bacteriol. Symp. Suppl.*, **73**, 103S–114S.
Dainty, R. H., Edwards, R. A., Hibbard, C. M. and Marnewick, J. J. (1989) Volatile compounds associated with microbial growth on normal and high pH beef stored at chill temperatures. *J. Appl. Bacteriol.*, **66**, 281–289.
Degn, H. (1992) Membrane inlet mass spectrometry in pure and applied microbiology. *J. Microbiol. Methods*, **15**, 185–197.
Degn, H., Cox, R. P. and Lloyd, D. (1985) Continuous measurement of dissolved gases in biochemical systems with the quadrupole mass spectrometer. *Methods Biochem. Anal.*, **31**, 165–194.
Degn, H., Bohátka, S. and Lloyd, D. (1992) Enzyme activity in organic solvent as a function of water activity determined by membrane inlet mass spectrometry. *Biotechnol. Techn.*, **6**, 161–164.
Doerner, P., Lehmann, J., Piehl, H. and Megnet, R. (1982) Process analysis of the acetone-butanol fermentation by quadrupole mass spectrometry. *Biotechnol. Techn,*, **4**, 557–562.
Edwards, R. A. and Dainty, R. H. (1987) Volatile compounds associated with the spoilage of normal and high pH vacuum-packed pork. *J. Sci. Food Agric.*, **38**, 57–66.

Edwards, R. A., Dainty, R. H. and Hibbard, C.M. (1987) Volatile compounds produced by meat pseudomonads and related reference strains during growth on beef stored in air at chill temperatures. *J. Appl. Bacteriol.*, **62**, 403–412.

Fenn, J. B., Mann, M., Meng, C. K., Wong, S. F. and Whitehouse, C. M. (1989) Electrospray ionization for mass spectrometry of large biomolecules. *Science*, **246**, 64–71.

Finlay, E. M. H., Games, D. E., Startin, J. R. and Gilbert, J. (1986) Screening, confirmation and quantitation of sulphonamide residues in pig kidney by tandem mass spectrometry of crude extracts. *Biomed. Environ. Mass Spectrom.*, **13**, 633–639.

Gilbert, J. (1987) (ed.) *Applications of Mass Spectrometry in the Food Sciences*. Elsevier Applied Science, Amsterdam.

Grotemeyer, J. and Schlag, E. (1988) Multiphoton-ionization mass spectrometry (MUPI-MS). *Angew. Chem. Intermat. Edn Engl.*, **27**, 447–459.

Gutteridge, C. S. and Puckey, D. J. (1982) Discrimination of some Gram-negative bacteria by direct probe mass spectrometry. *J. Gen. Microbiol.*, **128**, 721–730.

Hayward, M. J., Kotiaho, T., Lister, A. K., Cooks, R. G., Austin, G. D., Narayan, R. and Tsao, G. T. (1990). On-line monitoring of bioreactions of *Bacillus polymyxa* and *Klebsiella oxytoca* by membrane introduction tandem mass spectrometry with flow injection analysis sampling. *Analyt. Chem.*, **62**, 1798–1804.

Heinzle, E., Furukawa, K., Dunn, I. J. and Bourne, J. R. (1983) Experimental methods for on-line mass spectrometry in fermentation technology. *Biotechnology*, **1**, 181–188.

Henis, J. M. S. and Tripodi, M. K. (1983) The developing technology of gas separating membranes. *Science*, **220**, 11–17.

Heresztyn, T. (1986) Metabolism of volatile phenolic compounds from hydroxycinnamic acids by *Brettanomyces* yeast. *Arch. Microbiol.*, **146**, 96–98.

Hunt, D. F., Shabanowitz, J., Botz, F. K. and Brent, D. A. (1977) Chemical ionization mass spectrometry of salts and thermally labile organics with field desorption emitters as solids probes. *Analyt. Chem.*, **49**, 1160–1163.

Jardine, I. (1990) Electrospray ionization mass spectrometry of biomolecules. *Nature*, **345**, 747–749.

Jarvis, B. (1993) The chemistry and microbiology of cider making. In *Encyclopaedia of Food Science, Food Technology and Nutrition* (eds. R. Macrae, R. Robinson and M. Sadler). Academic Press, London, Vol. 2, pp.984–989.

Johnson, J. V., Yost, R. A., Kelley, P. E. and Bradford, D. C. (1990) Tandem-in-space and tandem-in-time mass spectrometry: triple quadrupoles and quadrupole ion traps. *Analyt. Chem.*, **62**, 2162–2172.

Jungclas, L., Schmidt, L., Fritsch, H.-W. and Köhl, P. (1991) ^{252}Cf-plasma desorption MS. *Internat. Lab.*, 25–30.

Karas, M. and Hillenkamp, F. (1988) Laser desorption ionization of proteins with molecular masses exceeding 10000 Daltons. *Analyt. Chem.*, **60**, 2299–2301.

Koppenaal, D. W. (1988) Atomic mass spectrometry. *Analyt. Chem.*, **60**, 113R–131R.

Kotiaho, T., Lauritsen, F. R., Choudhury, T. R., Cooks, R. G. and Tsao, G. T. (1991). Membrane introduction mass spectrometry. *Analyt. Chem.*, **63**, 875A–883A.

Lauritsen, F. R. (1990) A new membrane inlet for on-line monitoring of dissolved volatile organic compounds with mass spectrometry. *Internat. Mass Spectrom. Ion Proc.*, **95** 259–268.

Lauritsen, F. R. and Lloyd, D. (1993) Direct detection of volatile metabolites produced by microorganisms using membrane inlet mass spectrometry. In *Mass Spectrometry for the Identification of Microorganisms* (ed. Fenselau, C.). American Chemical Society (in press).

Lauritsen, F. R., Nielsen, L. T., Degn, H., Lloyd, D. and Bohátka, S. (1991) Identification of dissolved volatile metabolites in microbial cultures by membrane inlet tandem mass spectrometry. *Biol. Mass Spectrom.* **20**, 253–258.

Lauritsen, F. R., Kotiaho, T., Choudhury, T. K. and Cooks, R. G. (1992a). Direct detection and identification of volatile organic compounds dissolved in organic solvents by reversed-phase membrane introduction tandem mass spectrometry. *Analyt. Chem.* **64**, 1205–1211.

Lauritsen, F. R., Choudhury, T. K., Dejarme, L. E. and Cooks, R. G. (1992b). Microporous membrane introduction mass spectrometry with solvent chemical ionization and glow discharge for the direct detection of volatile organic compounds in aqueous solution. *Analyt. Chim. Acta* (in press).

Ledingham, K. W. D. and Singhal, R. P. (1991) Assessment of titanium-sapphire lasers and optical paramagnetic oscillators as sources of variable wavelength for resonant ionization mass spectrometry. *J. Analyt. Atom. Spectrom.*, **6**, 73–77.

Lee, E. D., Mück, W., Henion, J. D. and Covey, T. R. (1989) Liquid junction coupling for capillary zone electrophoresis/ion spray mass spectrometry. *Biomed. Environ. Mass Spectrom.*, **18**, 844–850.

Lindner, B. and Seydel, U. (1983) Mass spectrometric analysis of drug-induced changes in Na^+ and K^+contents of single bacterial cells. *J. Gen. Microbiol.*, **129**, 51–55.

Lister, A. K., Wood, K. V., Cooks, R. G. and Noon, K. R. (1989) Direct detection of organic compounds in water at parts-per billion levels using a simple membrane probe and quadrupole ion trap. *Biomed. Environ. Mass Spectrom.*, **18**, 1063–1070.

Lloyd, D. and James, C. J. (1987) The Pasteur effect in yeasts: mass spectrometric monitoring of oxygen uptake and carbon dioxide and ethanol production. *FEMS Microbiol. Letters*, **42**, 27–31.

Lloyd, D. and Scott, R. I. (1983) Direct measurement of dissolved gases in microbiological systems using membrane inlet mass spectrometry. *J. Microbiol. Methods* **1**, 313–328.

Lloyd, D., Bohátka, S. and Szilágyi, J. (1985) Quadrupole mass spectrometry in the monitoring and control of fermentations. *Biosensors*, **1**, 179–212.

Lloyd, D., Davies, K. J. P. and Boddy, L. (1986) Mass spectrometry as an ecological tool for *in situ* measurement of dissolved gases in sediment systems. *FEMS Microbiol. Ecol.*, **38**, 11–17.

Lloyd, D., Ellis, J. E., Hillman, K. and Williams, A. G. (1992a) Membrane inlet mass spectrometry: probing the rumen ecosystem. *J. Appl. Bacteriol. Symp. Suppl.*, **73**, 155S–163S.

Lloyd, D., James, C. J. and Maitra, P. K. (1992b) Mutations in phosphofructokinases alter the control characteristics of glycolysis *in vivo* in *Saccharomyces cerevisiae*. *Yeast*, **8**, 291–301.

Loo, J. A., Usdeth, H. R. and Smith, R. D. (1989) Peptide and protein analysis by electrospray ionization-mass spectrometry and capillary electrophoresis-mass spectrometry. *Analyt Biochem.*, **179**, 404–412.

Loo, J. A., Edmonds, C. G. and Smith, R. D. (1990) Primary sequence information from intact proteins by electrospray ionization tandem mass spectrometry. *Science*, **248**, 201–204.

Matsuo, T., Matsuda, H., Katakuse, I., Wada, Y., Fujita, T. and Hayashi, A. (1981) Field desorption mass spectra of tryptic peptides of human haemoglobin chains. *Biomed. Mass Spectrom.*, **8**, 25–30.

McLafferty, F. W. (ed.) (1983) *Tandem Mass Spectrometry*. John Wiley and Sons, New York.

Posthumus, M. A., Kistemaker, P. G., Meuzelaar, H. L. C. and Ten Noever de Brauw, M. C. (1978) Laser desorption-mass spectrometry of polar non-volatile bio-organic molecules. *Analyt. Chem.*, **50**, 985–991.

Poulter, L., Green, B. N., Kaur, S. and Burlingame, A. L. (1990) The characterisation of native and recombinant proteins by electrospray mass spectrometry. In *Biological Mass Spectrometry* (eds. Burlingame, A. L. and McCloskey, J. A.). Elsevier, Amsterdam, pp. 119–128.

Prosser, S. J., Brookes, S. T., Linton, A. and Preston, T. (1991) Rapid, automated analysis of ^{13}C and ^{18}O of CO_2 in gas samples by continuous-flow, isotope, ratio mass spectrometry. *Biol. Mass Spectrom.*, **20**, 724–730.

Pungor, E., Jr., Perley, C.R., Cooney, C.L. and Weaver, J.C. (1980) Continuous monitoring of fermentation outlet gas using computer coupled mass spectrometry. *Biotechnol. Letters*, **2**, 409–419.

Ramsey, E. D., Games, D. E., Startin, J. R., Crews, C. and Gilbert, J. (1989) Detection of residues of chloramphenicol in crude extracts of fish and milk by tandem mass spectrometry. *Biomed. Environ. Mass Spectrom.*, **18**, 5–11.

Roepstorff, P. (1987) Plasma desorption mass spectrometry—a new tool in peptide and protein chemistry. *ESN-Europ. Spectrosc. News*, **73**, 18–23.

Roepstorff, P. (1992) High molecular mass analysis. *Abst. Nordic MS Conference*.

Rose, M. E. and Johnstone, R. A. W. (1982) *Mass Spectrometry for Chemists and Biochemists*. University Press, Cambridge.

Schindler, J. and Schmid, R. D. (1982) Fragrance or aroma chemicals—microbial synthesis and enzymatic transformation. *Process Biochem.*, **17**, 2–8.

Silfer, J. A., Engel, M. H., Macko, S. A. and Jumeau, E. J. (1991) Stable isotope analysis of amino acid enantiomers by conventional isotope ratio mass spectrometry and combined gas chromatography/isotope ratio mass spectrometry. *Analyt. Chem.*, **63**, 370–374.

Smith, R. D., Loo, J. A., Edmonds, C. G., Barinagar, C. J. and Udseth, H. R. (1990) New developments in biochemical mass spectrometry: electrospray ionization. *Analyt. Chem.*, **62**, 882–899.

Torgerson, D. F., Skowronski, R. P. and Macfarlane, R. D. (1974) New approach to the mass spectroscopy of non-volatile compounds. *Biochem. Biophys. Res. Commun.*, **60**, 616–621.

Toth, J. P. and Snyder, A. P. (1991) Detection of 1, 3, 5-triazine derivatives in crop samples by thermolysis-atmospheric pressure ionization tandem mass spectrometry. *Biol. Mass Spectrom.*, **20**, 70–74.

Watson, J. T., Schultz, G. A., Tecklenburg, Jr. R. E. and Allison, J. (1990) Renaissance of gas chromatography-time-of-flight mass spectrometry. *J. Chromatog.*, **518**, 283–295.

Weaver, J. C., Perley, C. R., Reames, F. M. and Cooney, C. L. (1980) Temporarily immobilized microrganisms: rapid measurements using a mass spectrometer. *Biotech. Letters*, **2**, 133–137.

Williams, A. A. and Tucknott, O. G. (1978) The volatile aroma components of fermented ciders: minor neutral components from the fermentation of sweet coppin apple juice. *J. Sci. Food Agric.*, **29**, 381–397.

Yost, R. A. and Enke, C. G. (1979) Triple quadrupole mass spectrometry for direct mixture analysis and structure elucidation. *Analyt. Chem.*, **51**, 1251A–1262A.

8 Chemical sensors

E. KRESS-ROGERS

8.1 Introduction

In recent years, novel chemical sensors for the food industry have become a commercial reality. Ten years ago, the Leatherhead Food Research Association (LFRA) embarked on a study of novel sensors being developed for other sectors and their potential adaptation to applications in the food industry. Sensors development was underway particularly in the defence and clinical sectors, in industrial safety and environmental protection and also in the automotive industry. The technologies employed included microelectronics, optoelectronics and biotechnology. The LFRAs technology transfer study (Kress-Rogers, 1985) proposed the development of, for example, solid state pH sensors for on-line measurements in foods based on field effect transistor devices (ISFET), of direct insertion probes for the assessment of food freshness based on biosensors and of immunosensors for bacterial and fungal toxins in foods. At the time, these suggestions were regarded as futuristic by many in the food industry.

Now, a commercial ISFET pH sensor specifically designed for selected food applications has already been routinely used in slaughterlines for over a year, a number of commercial ISFET pH probes designed for the QC (quality control) laboratory have appeared on the market over the same period, and more ISFET-based instruments are due to appear shortly (see section 8.3). A prototype knife type biosensor array for the assessment of meat freshness has been developed in the United Kingdom and proved successful in trials. Other biosensors have been tested in the Japanese food industry. Instrumentation systems for the on-line and at-line measurement of chemical, physical and microbial properties are being developed for applications in food quality assurance by a consortium of 40 companies (and government support) in Japan. A number of biosensor developments for food applications are being undertaken collaboratively by groups in Germany, the United Kingdom and Japan; active chemical sensors development is also being pursued in other countries, although generally less food oriented.

Optical immunosensors technology (section 8.6) has progressed in recent years and can now provide much better sensitivity. Commercial immunosensor applications for food are currently confined to benchtop instruments in research laboratories, but these may well provide information that could smooth the path for the development of immunosensors for QC applications.

The knife-type meat freshness probe (section 8.7) is an example of the ultrarapid assessment of food freshness using the marker (or indicator) concept. This encompasses the identification of a food property that is indicative of a complex condition such as freshness or ripeness, followed by the development of a sensor for the measurement of that food property. A number of biosensor combinations for fish and meat freshness have been explored, particularly in Japan where these foods are often consumed raw, but also in other countries. Possible applications of the marker concept using other sensor types have been investigated, for example, the application of gas sensors to the assessment of MAP (modified atmosphere packed) foods. A particularly successful application of the marker principle is a novel monitor for the heat induced deterioration of frying oil in use. The output of this sensor correlates with a chemical reference assay but is actually based on the measurement of a physical property. This instrument is now under commercial development and is due to be released for industrial trials shortly.

Another active area in chemical sensors research and development is that of gas sensor arrays for the recognition of an aroma, for example, of a particular coffee or whisky type (section 8.4). Ultimately, this approach aims to take on the function of taste panels in selected applications. The use of sophisticated data processing techniques is essential in the evaluation of the outputs from the sensor arrays used here. Both conventional pattern recognition techniques and neural networks are being used. Neural networks are also used to evaluate signals from combinations of sensors based on unrelated technologies so as to find correlations with properties of interest. Another new approach in making the best of existing sensors is the use of fuzzy logic techniques. In addition, robotic sampling and conditioning help in applying non-optimised sensors on-line (section 8.8)

Thanks to advances in all these areas, the development and application of sensors for the food industry is now a topic of increasing practical importance. The development of a practical industrial instrument based on a new technology and moreover for a new application sector (namely the food industry) is full of potential pitfalls and requires many years of research and development before even the marketing aspects can be properly assessed. It is therefore a process that often appears painfully slow, leaving potential users frustrated and loosing interest. Once complete, the development of instruments based on new sensors can, however, provide considerable savings by allowing an early recognition of any deviation from the target properties during processing and thereby the avoidance of re-work or disposal. Process control based on suitable sensors allows an efficient use of labour, raw materials and energy resources. Many applications of chemical sensors in the food industry are aimed at the ensurance of quality and safety of the food product and on compliance with labelling regulations, and all these aspects will gain in importance in the coming years.

8.2 Sensor types

Chemical sensors will be defined here as sensors that determine a chemical variable such as the pH value or the concentration of inorganic ions such as sodium, organic acids such as lactic acid, compounds such as glucose or groups of compounds such as alcohols or specified agricultural residues or fungal toxins. The sensor types include firstly those based on interaction of an inorganic catalyst with the sample or on an ion exchange or a non-biological redox interaction. This group will be referred to as chemosensors here. The second group encompasses sensors based on an enzyme–substrate interaction with the sample and these will be labelled biosensors. Thirdly, chemical sensors can be based on an affinity binding or immunoreaction to the target analyte and these will be termed immunosensors. Many other definitions exist particularly for the term biosensor which is sometimes used to describe any sensor or even a tracer that can be introduced into a sample of biological origin. Narrower definitions are preferred here to distinguish sensor groups with quite distinct practical characteristics.

A sensor, unlike an indicator paper, has to be capable of giving a continuous signal to an electrical, optical or acoustic display device. A chemical sensor should, in general, have a reversible interaction with the target analyte. Many sensing heads are, nevertheless, designed as low cost disposable devices in order to avoid maintenance and hygiene problems. A more complex situation is encountered with the immunosensor. In contrast to the immunoassay, it can be configured to allow a continuous observation of the antigen–antibody binding process and thereby provide information on the binding kinetics as well as on the concentration of sample antigen. The immunosensing device can, in principle, be regenerated after a positive reading but in many instances this will lead to a deterioration of the characteristics. Often, the immunosensor is intended to trigger an alarm when a toxin, for example, is present. In such a case, the immunosensor can give a continuous signal until the alarm condition is reached. Immunosensors employing reagent delivery systems have been developed recently.

In all chemical sensor types, the primary signal from the interaction of the sensing surface with the analyte is transduced by a base device which carries the sensitising layer. The base device responds to electrical, optical, thermal, acoustic or mechanical properties of the sensing surface or to a secondary analyte generated by the interaction. Typically, it provides an electrical or optical output which can be displayed for reading by the user or transmitted to a process control system.

8.3 Microelectronic sensors for pH, acidity and ions

The pH value of food products is an important factor in their stability towards microbial spoilage and to chemical reactions. In food processing, the control of the pH value of the process stream is important for its relationship to the pH value of the final product, and also for the management of processes such as sugar inversion, fermentation or meat conditioning. The pH value also needs to be monitored for processing waste waters to avoid increased water treatment costs.

The pH value of food products is one of the main variables that would typically be measured as part of a hazard analysis critical control point HACCP programme. The pH value, the water activity and preservative concentration are essential factors in determining whether a food will support the growth of a particular group of microbes under ambient or chill conditions.

Unfortunately, on-line measurements of the pH during the processing operation have been hindered by the disadvantages of conventional glass membrane pH probes. In particular, the fragile glass membrane can represent a foreign body hazard, electrode function is often impaired by fouling of the glass frit liquid junction in the reference electrode, and maintenance is time consuming. Nevertheless, glass membrane pH probes were used in food process streams because the potential hazard from pathogens in the case of a pH above the limit value for the product was considered more severe than that from glass fragments in the case of probe damage.

To help avoid this dilemma, solid state pH probes together with solid state reference probes are becoming available. Such a solid state probe system is free of fragile membranes, filling solutions and glass frits. To date, the best performing device in this category is the pH ISFET/REFET system. The ion-selective field effect transistor (ISFET) is a microelectronic device and has been developed for the measurement of pH and of ions such as sodium, potassium, calcium and nitrate (Bergveld, 1970, 1988; Sibbald *et al.*, 1984; Van den Berg *et al.*, 1991; Klein, 1991). It is based on the FET device that is found as a pre-amplifying element in integrated circuits. The FET has been modified by replacing the metal gate film with an ion-selective film, and the gate contact has been replaced with a pseudoreference electrode (Figure 8.1). The pH ISFET device can be combined with a REFET device that is similar to the ISFET but has been designed to exhibit a very weak response to pH. The differential output is then measured. The modern commercial probes (Figure 8.2) overcome the

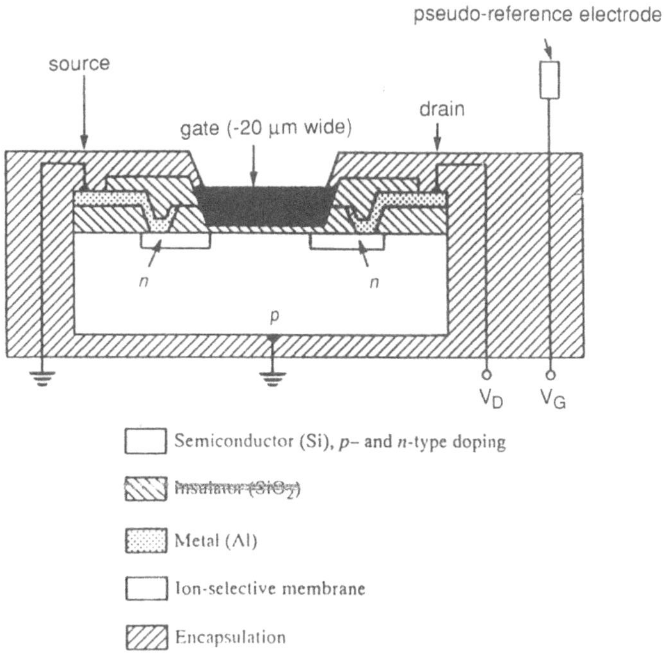

Figure 8.1 Schematic diagram of the ISFET. From Kress-Rogers (1993a).

drift problems that were observed with the early pH ISFET reference systems.

Maintenance is simplified by the absence of a fragile membrane or of a glass frit with a tendency to foul. A commercial pH ISFET/REFET system by Sensoptic has been successfully used in slaughterhouses now for over a year to monitor the post-mortem pH change in carcasses. The probe has been configured as a robust stainless steel body stab probe for insertion into carcasses or through tough sausage casings. In trials with the probe, it was shown that by measuring at two specified times within the first day, conditions such as pale, soft, exudative (PSE) and dark, firm, dry (DFD) (which indicate lower meat quality) can be determined (Den Reijer, 1990; Eikelenboom, 1990a,b). The probe can be cleaned by brushing with alcohol on a tooth brush. Corrosive cleaning solutions should not be in contact with the probe for prolonged periods. Temperature stability is limited by the adhesive used in the present probe which operates up to 70°C. This limit could be raised in future probes. Drift is specified as 0.05 pH units per 24 hours. Several companies, for example UNIFET, now offer ISFET pH probes intended for the laboratory. The absence of the

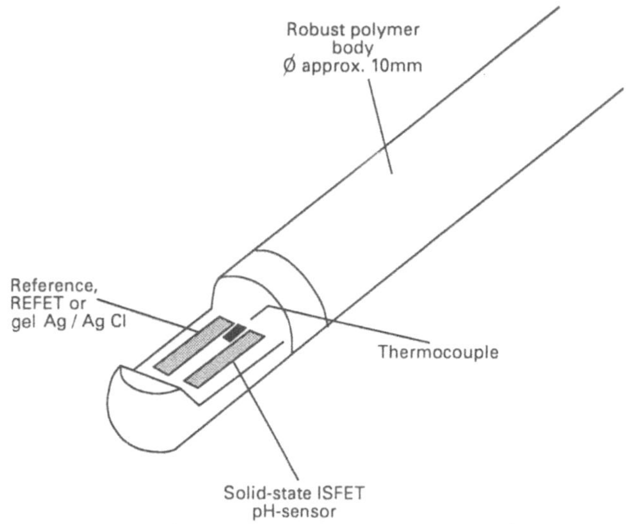

Figure 8.2 Commercial ISFET probe for liquid and semi-solid foods. The solid state pH sensor is mounted in a robust break-resistant polymer body, eliminating glass membranes, glass frit, liquid junctions and glass bodies. The all-solid construction can be cleaned with a brush and provides robustness, eliminates glass fragment hazards and is clog-resistant. Completely submersible probes are available. In contrast to a conventional pH glass membrane electrode, the ISFET probe can be stored dry. Due to this and the absence of filling solutions, routine maintenance is eliminated. Shown is a configuration for liquid and semi-solid materials. In addition, a robust stainless steel stab probe that can be inserted into meat carcases and meat products, has been evaluated in the meat industry.

glass membrane and glass frit and thus of the foreign body hazard and maintenance problems is an advantage not only on-line but is also preferable for at-line measurements in the production area. In contrast to the conventional glass membrane electrode, the ISFET probe can be stored dry. Another advantage of the ISFET/REFET system is the fast response and this can be used to advantage in flow injection systems. Furthermore, a good signal-to-noise ratio can be achieved without heavily shielded cables (after all the ISFET originates from a pre-amplifier device!).

A lightly built inexpensive probe with a small sensing tip is made by Microsens. Many foods are inhomogenous in their chemical composition, and a few studies have been carried out on the effect of small regions of a higher pH within an otherwise low pH food on microbial growth. Such investigations will be facilitated by pH ISFET probes with fine tip configurations.

The small sensing element and very compact instrument size also lend themselves to the construction of multi-sensor arrays, for example for pH, sodium and potassium. The sensing elements, as well as the associated instrument, complete with miniaturised pumps for sample and calibration liquids, can be miniaturised as pocket instruments. So far these have only been developed as laboratory prototypes intended eventually for clinical use. Other multi-ion ISFET configurations have been developed for agricultural applications.

Not only is the pH value of interest in the food industry but also the titratable acidity. While the pH is relevant primarily for microbial growth and the rate of chemical reaction, the acidity is more important for the taste. At-line automatic titrators exist already, based on conventional methods. In addition, a true on-line measurement of acidity is now a possibility (pending further development) with a coulometric microtitrator device based on an ISFET together with a microactuator electrode on the same chip (Van der Schoot and Bergveld, 1985; Kress-Rogers, 1986). Moreover, for liquids of low viscosity, it is possible continuously to display a complete titration curve with all the associated information on relative acid concentrations. This is achieved by constructing an array of microtitrator devices within a special flow cell. For more viscous samples, a single microtitrator in a dipstick configuration is possible instead (Olthuis *et al.*, 1989).

Further details on ISFETs for the measurement of pH, ions and acidity are given by Kress-Rogers (1993a). Alternative approaches for solid state pH sensors are also discussed there. A briefer update on solid state pH sensors can be found in Kress-Rogers (1991).

8.4 Sensor arrays for aroma and indicator volatiles

Sensors for gases such as methane or carbon monoxide were initially developed for application in the mining and automotive industry, based on metal oxide (MeOx) semiconductors such as tin oxide or zinc oxide. A well known representative is the Taguchi sensor of the Figaro campany in Japan consisting of a ceramic former on which tin oxide and an electrode for heating the MeOx are deposited.

The conductivity of simple resistor devices made from these materials changes in response to a wide range of oxidising and reducing gases and volatiles including, for example, alcohols (Kohl, 1989). Their selectivity for a particular set of gases or volatiles can, to some extent be tailored by doping and tempering procedures at the time of preparing the sensor, and later by choosing the temperature at which the sensor is kept during operation. In general, however, the selectivity of this sensor type is broad and it is not as sensitive as certain other gas sensor types such as the chemically sensitive field effect transistor (CHEMFET). Variability between devices of the same batch is also more pronounced for the MeOx semiconductor resistor. (The usual MeOx gas sensor is a simple resistor structure in contrast to metal–oxide–semi-conductor (MOS) devices such as the field effect transistor (FET). The MOS device consists of a semiconductor substrate carrying an insulating oxide layer and a metal film. With a catalytic metal film as the top layer, the MOSFET device becomes the CHEMFET gas sensor.)

Beyond the straightforward application in combustion control and mining safety, the MeOx gas sensor has found a further range of applications. Here, one uses the broad specificity to advantage in a manner analogous to the principle upon which a human nose operates. The human nose can distinguish a vast range of gases, volatiles and aromas. It does this not by providing a specific receptor for each compound, but by combining the information from receptors with broad overlapping specificities (Persaud and Dodd, 1982). (A similar approach is also used in the human eye in colour perception.)

A simple system imitating this approach is an array of identical MeOx sensors each kept at a different temperature. One can then use simple ratios or complex pattern analysis techniques (either conventional or neural network-based) to analyse the output of this sensor array and to find correlations with the composition of gas mixtures (Gardner, 1991). A commercial instrument of this type is on the market, using a simple design (chamber with five MeOx devices at different temperatures, simple signal evaluation). This has been used primarily in applications such as the curing of epoxy resins, but the manufacturers also report the monitoring of coffee roasting operations.

Distinction between the aromas of different food varieties has also been suggested and more complex signal evaluation procedures are generally used here. There is no reason, however, why the importance that this gas sensor array attaches to a particular compound in the gas/volatile mixture is the same as the importance attached by the human nose. It is therefore advisable to use extensive test procedures before attempting to use such an approach in any application related to aroma assessment. In the tests, it is necessary to simulate all the conditions that might be encountered in use: type and concentration of gases and volatiles originating either from the product to be assessed or from other sources including air humidity.

It is likely that the application of gas sensor arrays to the detection of the end point of cooking, roasting and drying operations will be easier to set up than their application to the distinction of, say, fruit aroma. It is also possible that applications in the monitoring of food spoilage or the assessment of food freshness can be realised.

Another class of volatile sensors are the conducting polymer gas sensors based particularly on polypyrrole or polyaniline. Again, these have a broad specificity that can be tailored to some extent. They have been configured in the form of arrays of up to fifty similar devices combined with pattern recognition techniques (Persaud and Pelosi, 1992). A commercial development of odour sensors based on these devices is planned. Their long term stability still needs to be assessed once they are more widely available.

Highly sensitive gas sensors with a narrow specificity for NO_2, CI_2 and O_2 are based on metal phthalocyanine films. The CHEMFET sensors are also highly sensitive and specific. They respond to the concentration of gases such as hydrogen, hydrogen sulphide, methane, ammonia and ethylene depending on the sensitising film chosen (Sundgren *et al.*, 1990). The latter consists of a metal or metal alloy film which is incorporated into the gate area of the device. CHEMFETs are manufactured in planar silicon technology which provides highly reproducible devices. Cost per device decreases strongly with the number of identical devices that can be produced per year. A feasibility study has been carried out on the application of CHEMFET sensors in the monitoring of spoilage in modified atmosphere packaged foods (see section 8.7). It is also worth investigating the possibility of applications in the monitoring of fruit ripening.

Where the highly specific determination of volatiles such as particular alcohols, amines or alcohols is needed, biosensors should be considered. These are described in sections 8.5 and 8.6.

8.5 Biosensors based on amperometric, potentiometric and thermometric devices

The sensors discussed above are based on progress in the field of semi-conductor technology. A further impetus to the development of chemical sensors has come from advances in biotechnology. The first biosensor was constructed by Updike and Hicks in 1967. This consisted of a Clark-type oxygen electrode on which a membrane was held with an O-ring. The membrane served to retain the isolated and purified enzyme glucose oxidase. When this probe was immersed in an aqueous buffer containing glucose, it responded by indicating a decrease in oxygen concentration caused by the enzyme-catalysed oxidation of the glucose. Subsequently, other base devices including peroxide and pH probes were used with the same enzyme.

Later, some workers replaced the conventional Clark-type electrodes with micro-Clark electrodes constructed in planar silicon technology (both amperometric devices), others used ISFET pH sensors in place of conventional pH electrodes (both labelled as potentiometric devices); yet others used CHEMFET gas sensors as base devices (Caras and Janata, 1980; Gotoh *et al.*, 1986; Danielsson and Winquist, 1989; Koudelka *et al.*, 1989; Karube *et al.*, 1990; Kimura and Kuriyama, 1990; Brand *et al.*, 1991).

Another approach was taken by workers who developed amperometric biosensors that did not incorporate a gas probe and could operate without ambient oxygen. They achieved this by combining the oxidase enzyme with a compound that replaced oxygen in the enzyme–substrate interaction (Figure 8.3). For this purpose, dimethylferrocene and related compounds (Cass *et al.*, 1984) were found to be particularly suitable. Further electron mediators have since been developed (Turner *et al.*, 1987) partly to overcome the restriction on the use of ferrocene by patents, partly to apply the mediator approach to non-oxidase enzymes. The enzyme and mediator can be immobilised together onto porous carbon which serves as electrode material. This type of biosensor became known initially as a direct biosensor (a label soon disputed by the followers of the conducting salt electrode approach which was considered as more direct) and later as amperometric mediated enzyme electrode (AMEE). Biosensors based on this device have been developed particularly by A.P.F. Turner and his colleagues at Cranfield Biotechnology Centre.

A lively debate on the merits of these differing biosensor approaches has taken place. The early potentiometric biosensors were criticised for their poor signal-to-noise characteristics, but this does not apply to the ISFET-based biosensors (known as ENFET). The application of the ENFET principle has, however, been held back by the slow initial progress with commercial ISFET devices. The recent emergence of a number of commercial ISFET pH sensors could change this.

The AMEE is the basis of an elegant commercial instrument that is used by diabetic patients to measure their blood glucose concentration (The EXACTECH by Medisense). A drop of blood is deposited onto a paper strip which carries a disposable AMEE device. The strip is inserted into a pen- or card-shaped instrument with a digital display. The instrument is switched on and, thirty seconds later, displays the glucose content. The electrode strip is then disposed of. An advantage of this approach is that a calibration for each AMEE strip is not needed. Instead each batch of, for example, fifty AMEE strips is provided with a calibrator strip which is used just once when the pack is started.

There has been much research aimed at developing the AMEE into an implantable continuous *in vivo* monitor, but as yet many problems remain

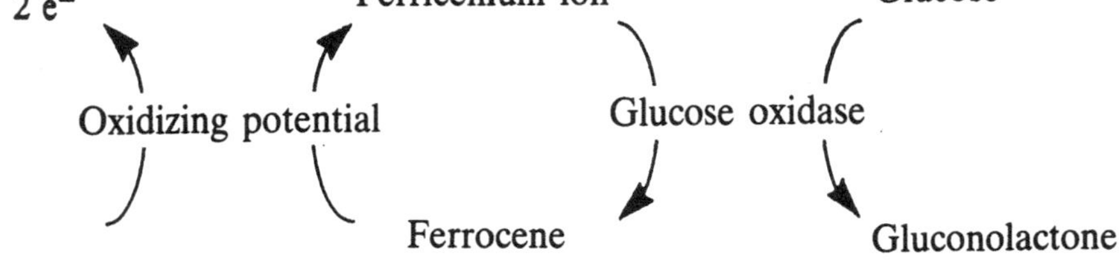

Figure 8.3 Principle of the amperometric mediated enzyme electrode (AMEE).

to be solved here particularly in relation to the gradual loss of mediator from the sensor. Instead, another device type developed from the first generation biosensors has been more prominent so far in *in vivo* trials (Mascini, 1989). This device is the amperometric indirect biosensor, based on a needle-type peroxide probe (or a needle-type oxygen probe). The performance of this device has progressed primarily through advances in membrane technology immobilisation techniques. In the amperometric indirect dual membrane enzyme electrode (abbreviated below as AID-MEE), two membranes (MacDonnell and Vadgama, 1989) tailored individually for specific tasks such as control of analytical range and reduction in electrochemical interference and biofouling of the sensor are incorporated (Figure 8.4). The AIDMEE is insensitive to oxygen concentration down to very low oxygen pressures and thus a major disadvantage of the early indirect biosensors has been addressed. In the absence of oxygen, aeration of the device is possible.

An advantage compared with the AMEE is the possibility of eliminating ascorbate interference without a notable increase in response time. The AMEE, on the other hand, still has the advantage of having been commercially proven as a device that can be manufactured with a high

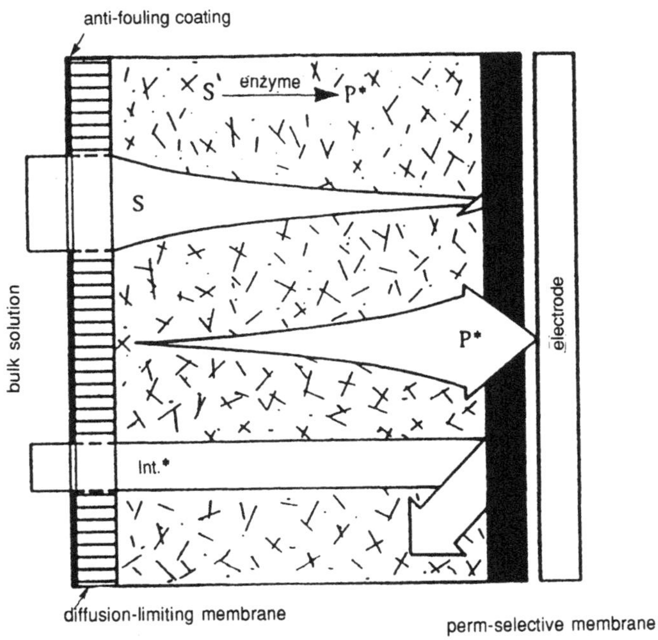

Figure 8.4 Amperometric indirect dual membrane enzyme electrode (AIDMEE). Schematic design. S = substrate, P* = product, Int.* = interferent, * = electromagnetically active. From Kress-Rogers (1993a). After MacDonnell and Vadgma (1989).

device-to device reproducibility, at least for single use. For continuous use, AMEE devices at present require frequent re-calibration. To overcome this, self-calibration features have been built into AMEE probes for continuous operation in molasses (Bradley and Schmid, 1991).

For foods, an important consideration is the cost of developing a range of biosensors for different analytes at different concentrations in a variety of chemical and physical backgrounds. Here, the AIDMEE can provide a potentially lower adaptation cost by using a simpler electrochemical system, and tailoring of device characteristics by the dual membrane system (Vadgama, 1990). A successful development based on an AMEE glucose sensor array is that of a knife-type meat freshness probe (see section 8.7). Many indirect biosensors have been tested in the Japanese food industry. They make use particularly of robotic sampling and conditioning techniques and of flow injection techniques in preference to full biosensor optimisation (see section 8.8).

For certain analytes several enzymes and sometimes co-factors are needed in the biosensor types described above. For such analytes, another biosensor type can be constructed with a single enzyme only. This was initially the enzyme sensitised thermistor (ENTHERM). More recently, more reproducible heat sensitive devices such as the thermopile have been employed. With the ENTHERM, sucrose for example can be determined by immobilising invertase on the heat sensitive device. The ENTHERM is used in conjunction with a reference device free of active enzyme to compensate for temperature variations in the sample. With the earlier thermistor based devices, a water bath was additionally needed to achieve high sensitivities, but it may be possible to obviate the water bath with the more reproducible thermopile as a base and reference device (Danielsson, 1990; Muehlbauer *et al.*, 1990).

8.6 Chemically sensitive optical and acoustic devices

An alternative to the microelectronic pH and ion sensors described in section 8.3 are the fibre-optic chemical sensors. Fibre-optic probes (FOP) for the pH value and the concentration of oxygen and carbon dioxide in blood are already routinely used in surgery. A FOP for pH consists of an optical fibre section which carries a pH-sensitive dye. In some laboratory prototypes, the dye is simply adsorbed onto porous resin spheres held onto the fibre tip by a membrane but this design is liable to drift problems. Good stability for pH values around 7, as in the clinical blood pH probes, can be achieved by dye entrapment in acrylamide gel. More recently, much progress has been made in the immobilisation of colorimetric reagents by covalent binding to polymer layers grown on the fibre-optic tip thus providing good stability for a wider range of conditions (Figure 8.5).

The FOP is connected to an optical instrument which illuminates the device through a connecting fibre and monitors the reflected intensity, typically at two or more wavelengths, to quantify the colour change in response to the analyte. Ambient light interference is avoided by using pulsed illumination and a matching detector. This general principle can be applied to the measurement not only of pH value but also of the concentration of a number of ions and gases. In principle, any colorimetric reaction for which a reversible dye exists, can be transformed in this way so as to obtain a quantitative and continuous measurement. FOP can also be turned into biosensors and immunosensors. Reagent delivery systems have been designed for continuous operation with non-reversible interactions as in immunosensors (Arnold and Meyerhoff, 1988; Arnold, 1990; Leiner and Wolfbeis, 1991; Walt *et al.*, 1991; Wolfbeis, 1991).

Advantages of the FOP approach include insensitivity to electromagnetic interference and mains noise, and safety in electrostatic or steamy

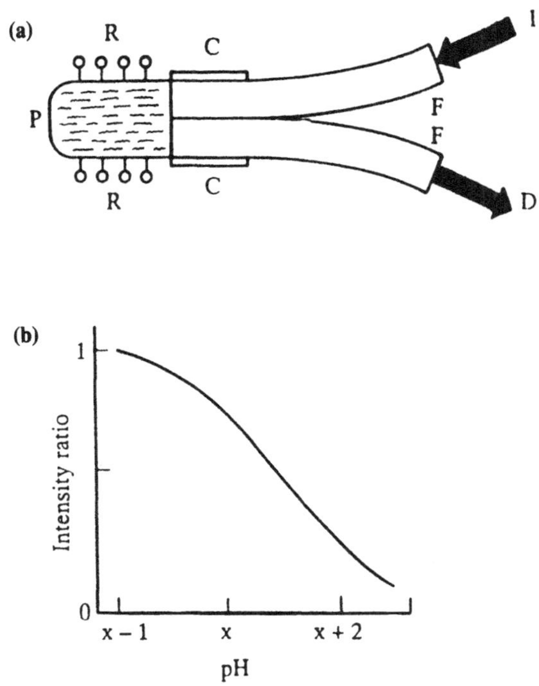

Figure 8.5 Fibre-optic pH sensor: (a) schematic diagram, (b) typical response. From Kress-Rogers (1993a).

environments. Also, the connecting optical fibre can be chosen to extend over several metres thus providing a long flexible probe that could, for example, be lowered into a tank or drain. A disadvantage for pH is the narrow range of typically 2 pH units for each dye, but this could be overcome by a fibre bundle carrying several dyes, for example. As in the case of ISFET pH and ion sensors, enzymes have been incorporated into FOP ion and gas probes to produce biosensors. In the food industry, fibre-optic temperature probes are already routinely used in industrial microwave ovens. In this environment they can operate without interfering with the performance of the oven and without suffering interference from the electromagnetic radiation.

While the FOP could be regarded as a carrier for an optical signal with an attached colourimetric reagent (admittedly in an often sophisticated form), the surface plasmon resonance (SPR) device is a true optical device. It allows the measurement of minute changes in the refractive index of a surface layer adjacent to the device. For an SPR device, there is an angle of incidence, where light is totally absorbed by the electrons in a thin metal layer that forms part of the device. The angle at which this occurs changes very sensitively with the refractive index in the immediate vicinity of the metal layer. The device can be constructed using either a prism (mainly for laboratory prototypes) or a grating configuration.

The first application of the SPR as a base device for chemical sensing (Figure 8.6) was by applying an absorbent film. In contact with certain gases, this film would swell and change its refractive index. This was used for the determination of halothane gas and was shown to be highly sensitive (Nylander *et al.*, 1982). Subsequently, the device was configured as an immunosensor by immobilising antibodies on the sensing surface (Liedberg *et al.*, 1983: Daniels *et al.*, 1988). An SPR based instrument is now available commercially in benchtop form (Pharmacia). In this form, it is intended as a research tool, allowing the monitoring of the antibody–antigen binding process in real time and thereby providing much more information than a conventional immunoassay. The binding rate in the initial and later stages of the binding process can be quantified, for example.

In the future, the development of more compact instruments dedicated to specific applications based on the SPR device is possible. Such devices would have the following advantages compared with the conventional immunoassay: rapid result (short incubation) and simple procedure (no steps to separate bound from unbound antigen). A design providing enhanced sensitivity has been described recently. Immunosensor applications in the food industry (Kress-Rogers and Turner, 1986) can be envisaged particularly for pathogens such as *Salmonella* and *Listeria*, for mould counts and for vitamin D. Other possible applications are for meat speciation and for the assay of mycotoxins.

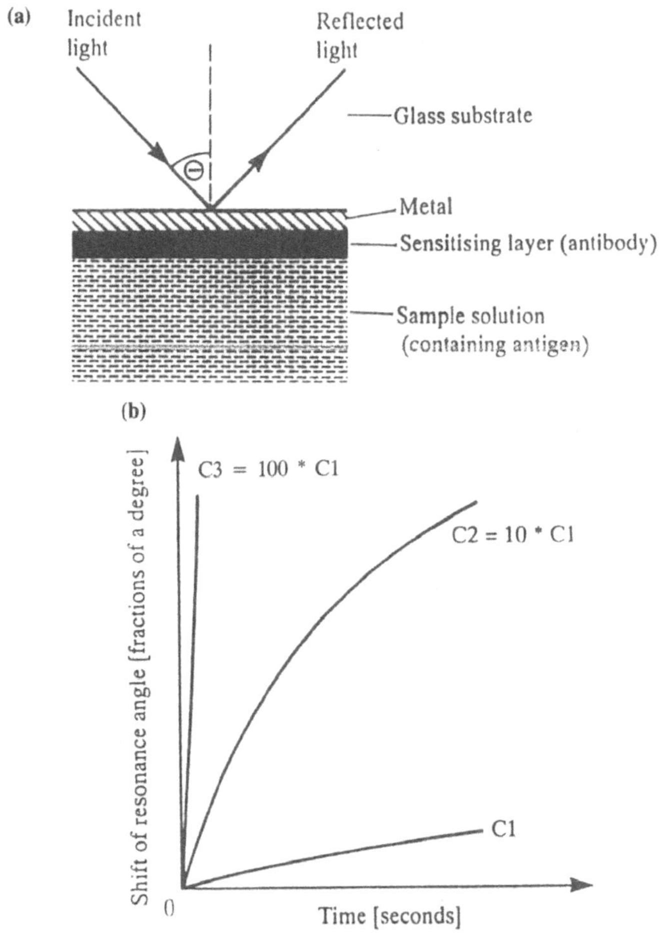

Figure 8.6 Surface plasmon resonance (SPR) device for the measurement of minute changes in the refractive index. (a) The device can be configured as a highly sensitive gas sensor or as an immunosensor by the immobilisation of sensitising layers on the device. From Kress-Rogers and Turner (1986). (b) In the immunosensor configuration, it is possible to monitor the antibody–antigen binding process directly without a prior incubation period unless the antigen concentration is very low (as for the concentration (C_1)). The sensitivity for kinetic monitoring can be enhanced with labelling techniques (see text). From Kress-Rogers (1993a). Adapted from Liedberg *et al.* (1983).

Immunosensors have also been constructed on the basis of total internal reflection (TIR) devices, also known as evanescent wave immunosensors (EVIs). When light enters a waveguide at a certain angle, it will be transmitted through the guide in a zig-zag path, being reflected from the upper and lower surface of the waveguide. In this manner, all the light is transmitted through the guide without losses to the adjacent medium (Figure 8.7). However, the light wave persists within a distance of the order of a wavelength into the other medium as an evanescent wave.

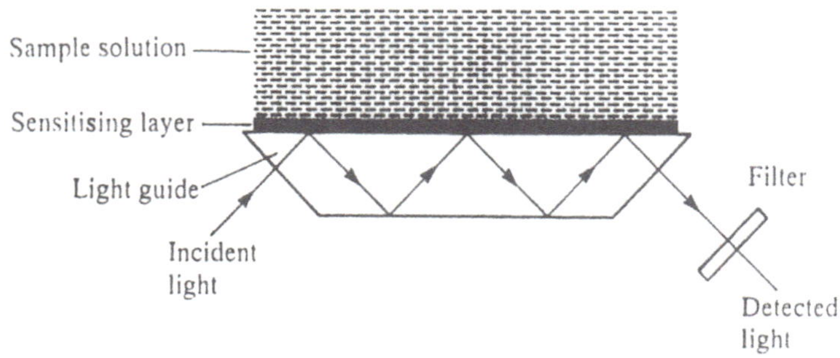

Figure 8.7 Total internal reflection device (TIR) makes use of the effect of analyte binding to an immobilised sensing layer, on the evanescent wave. From Kress-Rogers and Turner (1986).

Normally, this will not affect the transmission of light through the guide. However, if the refractive index of the surface layer changes or if centres of absorption are present there, then less light will be transmitted. This effect is used in the design of immunosensors by applying a sensitising antibody layer on the waveguide surface. Fluorescent labelling techniques can be used to advantage. The device has been configured, for example, as a capillary cell (Badley *et al.*, 1987) so that a defined amount of sample is automatically taken.

There are also two acoustic device types that have frequently been configured as immunosensors. The surface acoustic wave (SAW) device transmits and receives an acoustic wave that propagates in the surface region of a piezoelectric substrate. Wave propagation is influenced by changes in the density, elasticity and electrical conductivity of the surface layer. Again, the gas sorbent or antibody layer is immobilised on the surface of the device (Ballantine and Wohltjen, 1989; D'Amico and Verona, 1989). Operation in liquids has been problematic due to the strong damping of the transmitted signal when the surface is immersed in liquid. However, current research is addressing this problem.

The piezocrystal balance supports a bulk acoustic wave. It is used to measure minute changes in the mass of layers deposited onto it. This response to mass change is also used in its configurations as chemical sensor. Adsorbing films for gas detection, enzymes for a biosensor configuration and antibodies for operation as an immunosensor have all been used as sensitising layers and high sensitivities have often been achieved (Guilbault and Jordan, 1988; Muramatsu *et al.*, 1989; McCallum, 1989). A reference device is needed to compensate for changes in humidity and for the deposition of dust. (Further applications of piezoelectric devices and acoustic techniques particularly for biomass measurement are discussed by Clarke *et al.*, 1987).

8.7 Sensor applications employing the marker approach

An important aspect of quality assurance in the food industry is the assessment of the microbial status of foods. The presence of microbes on foods specified to be commercially sterile needs to be detected, a total load below a specified limit needs to be ascertained for other types of foods, and the absence of pathogens on foods or surfaces in the processing area must be ensured. For the latter application, immunoassays are used now and immunosensors could play a role in the future. For the determination of total microbial load, rapid methods have been developed. One of these is based on the monitoring of the impedance of a culture broth inoculated with the sample. Another is based on the determination (by a luminescent method) of bacterial ATP after separation from ATP endogenous to the food. The impedance method takes typically four or more hours depending on the microbial load in the sample. The ATP method takes approximately half to one hour depending on whether a separation step is needed. Both methods require an investment into a sophisticated instrument but provide the result much faster than the traditional plate count which takes several days.

A result within a minute or so can be achieved if the marker approach can be implemented for a particular type of food. Such an ultrarapid method can be developed by identifying a sample property which is indicative of the microbial load and then developing an instrument that can measure this property directly. An example for this approach is the development of a knife-type probe for the assessment of meat freshness. Bacteria at the surface of chilled meat initially (in the pre-spoilage period) consume the glucose present in the meat preferentially. When the glucose concentration has decreased substantially, the bacteria switch to the consumption of amino acids and begin to produce unpleasant amines at this stage. By measuring the degree of glucose depletion, there is therefore the possibility of obtaining an assessment of microbial activity at the pre-spoilage stage, when a measurement is of particular interest. As the microbial flora grow, the glucose concentration at the meat surface decreases and a glucose gradient into the depth of the meat is formed.

The resulting glucose concentration depth profile was determined for a model system by Gill (1976). Measurements at the LFRA confirmed the formation of this profile with microbial growth for lamb, beef and pork with its native bacterial flora. The study also provided data on the variation of the bulk glucose concentration with meat species and cut. The pH value at the surface and in the depth of the meat was also measured during the trial. As a biosensor principle for the development of a probe for the glucose profile, the AMEE was chosen since this had reached an advanced stage of development for blood which has some characteristics in common with meat juice. Based on the specifications from the LFRA study, an

AMEE glucose sensor array was developed at Cranfield Biotechnology Centre.

The resulting four-point glucose sensor array was tested in a trial with pork loins. The results of the trial (Figure 8.8) suggest that a fully developed prototype could provide a prediction of shelf life before organoleptic changes are observed. Such a measurement with a fully engineered prototype could provide results within minutes with a simple direct insertion measurement (Figure 8.9). This collaborative project was part of a programme undertaken for the Ministry of Agriculture, Fisheries and Food by the LFRA. A commercial investment will be needed to provide a commercial prototype suitable for extensive trials with meat samples representing different meat species, origin, cuts and slaughter-

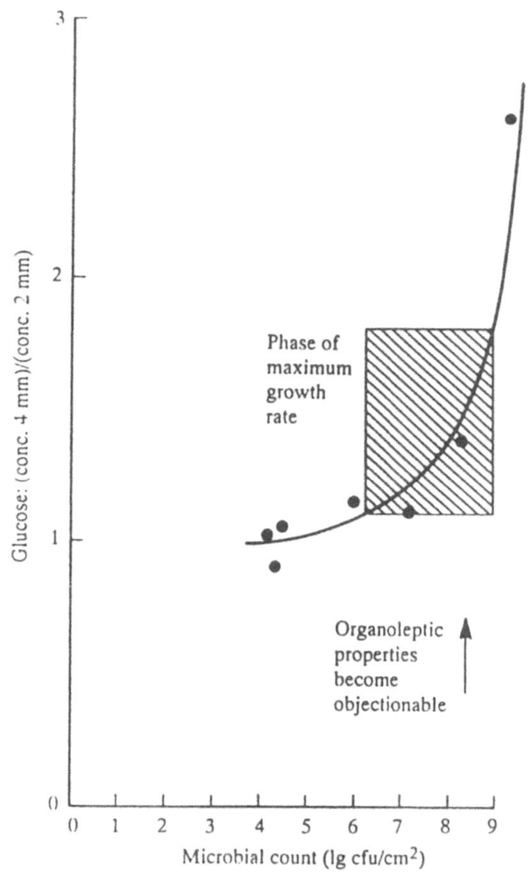

Figure 8.8 Trial with the glucose biosensor array in pork loins. Increase in the glucose concentration gradient with microbial growth. Each point in the graph represents an average over the four loins in the trial. From Kress-Rogers (1993b).

house conditions prior to probe release. More details can be found in Kress-Rogers (1993b) and in Kress-Rogers *et al.*, (1992). (The earlier stages of development are described in Kress-Rogers and D'Costa, 1986; Kress-Rogers *et al.*, 1988).

A number of biosensors for meat and fish freshness have been constructed in Japan, more recently also in collaboration with Germany. These include biosensors for amines such as cadaverine and putrescine for meat and fish (more recently also in combination with glucose sensors), and histamine for dark-fleshed fish. For very fresh fish to be eaten raw, hypoxanthine, inosine and ATP are measured (Watanabe *et al*, 1986; Karube and Tamiya, 1989).

For the assessment of fruit and vegetable ripening and conditioning, both the concentration of saccharides in the produce and, in some applications, that of ethylene in the headspace can be employed as indicators. The sugar concentration is not only relevant for the taste of a fresh or cooked fruit or vegetable, but is also important in determining the degree of browning in roasting and frying operations. The latter is due to the Maillard reaction between reducing sugars and amino acids at high temperatures. The measurement of glucose in produce extracts with the help of a commercial biosensor for healthcare applications has been investigated at the LFRA, and a procedure for a rapid sample preparation was found to make the application of this commercial instrument possible (Figure 8.10). A biosensor combination specifically intended for the measurement of saccharides in produce is under development in Japan and in Switzerland at present, a glucose AMEE has been used for the analysis of banana extracts (D'Costa *et al.*, 1988).

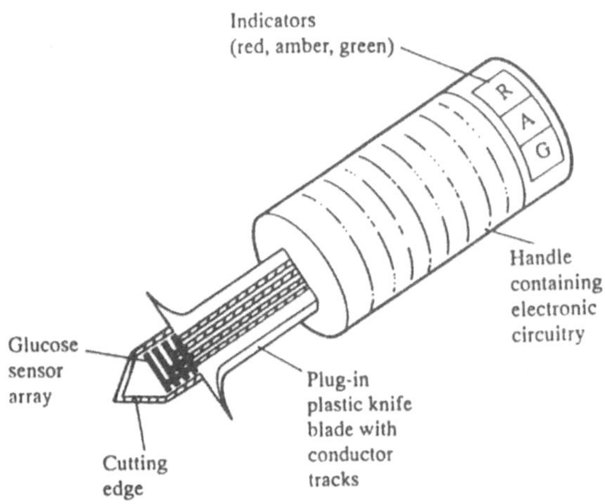

Figure 8.9 Proposed form of the instrument for the assessment of chilled meat by the glucose depletion profile. From Kress-Rogers (1993b).

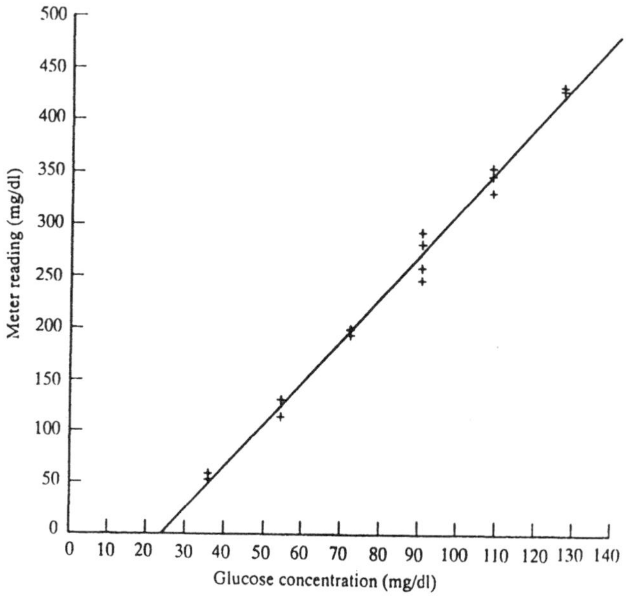

Figure 8.10 Characteristics of the ExacTech system for blood glucose measurement when challenged in non-blood samples. Each point on the graph corresponds to an individual disposable biosensor strip used without prior calibration. Calibration for assorted fruit juices after dilution (1:30) with a buffered solution. The reference measurements (*x*-axis) were taken with Dionex (HPLC) and Yellowsprings (enzyme electrode based analyser) instruments. From Kress-Rogers (1993a). Results reported earlier at LFRA Panel Meetings 1988/1989 by J.E. Sollars and E. Kress-Rogers.

A GC-MS study was carried out at the LFRA in order to identify suitable indicators of oxidative rancidity in the headspace of oils, fats and foods containing these. Certain aldehydes were found to be good indicators, namely hexanal and pentanal. If biosensors for these particular aldehydes were to be developed, a non-destructive continuous monitoring in storage vessels, for example, would be possible.

Volatiles are also produced in the spoilage of foods. Cleland *et al.* (1984) measured hydrogen emanating from bacterial cultures with a CHEMFET hydrogen sensor. A preliminary feasibility study was carried out at the LFRA to examine the possibility of monitoring the production of hydrogen in MAP foods non-invasively. A CHEMFET (courtesy of Thorn EMI) was placed onto the covering film of MAP raw meat and fish. With a slight modification to the sensing head, hydrogen production could be observed without breaching the gas-barrier film from the time of the best-before-date in some of the samples, later in others. The permeability to hydrogen of the polyethylene (PE)-nylon films tested in the study was found to be sufficiently reproducible across the film roll for such a measurement. Condensation does affect the permeability but many films are treated with anti-condensation coatings. Hydrogen was more readily produced on meat steak than on fish steak (probably due to higher glucose concentration in meat), and more readily on whole gutted mullet than on cod steak (more Enterobacteria in the gut flora) (Kress-Rogers, 1987). Further studies are needed to establish whether hydrogen is a suitable indicator. However, the general principle of a headspace measurement of an indicator gas or volatile (and in some cases even a measurement through a covering film) for spoilage assessment was demonstrated in the study.

Traces of alcohols including methanol and ethanol are indicators of the onset of both wanted and unwanted fermentation by yeasts. In higher concentrations, they are relevant both in fermentation control and in labelling and duty assessments. In general, biosensors for alcohols are more sensitive to methanol than to ethanol. The former is present in small quantities as a result of pectin decomposition.

For the assessment of frying oil deterioration, the development of a biosensor for an indicator compound was first considered at the LFRA. However, a study of literature and LFRA data showed that viscosity would be a more suitable indicator and could, moreover be measured *in situ* in the hot oil (Kress-Rogers, 1987/1988). Based on this study, a prototype was constructed at GEC Marconi Research Centre and evaluated at the LFRA. A special experimental procedure was developed for use with the early prototype to allow direct application in a catering fryer. The study showed that the output from the sensor (a vibrating tube viscometer with piezoelectric excitation) increased with the value for the reference assay POM (polymerised and oxidised matter by gas liquid chromatography, GLC). Moreover, this increase had a similar relationship to the increase in POM value for three very different oils (Kress-Rogers *et al.*, 1990). In this respect as well as in the *in situ* application, the sensor was shown to be

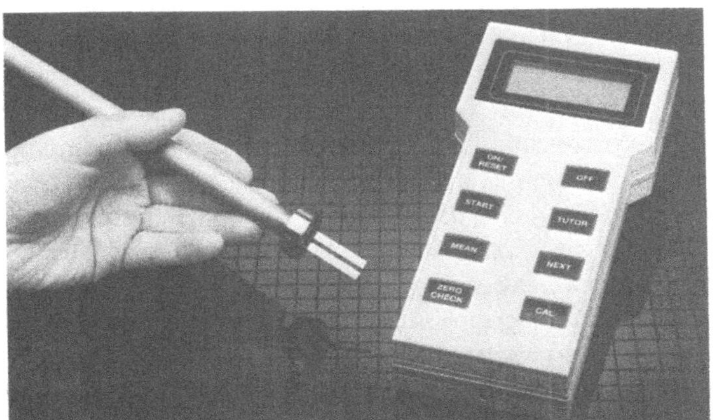

Figure 8.11 Pre-production model of frying oil monitor. The development of the instrument is described in Kress-Rogers (1993b). Photo by courtesy of GEC Sensors Ltd.

superior to the rapid test methods currently in use (See Kress-Rogers, 1993b for a description of this instrument development.)

This prototype has since been taken up by GEC Sensors Ltd for commercial development. A more advanced prototype has already been developed and evaluated. This is more easily applied than the earlier prototype, it provides results of enhanced reproducibility and has shown that a good resolution of the viscosity increase can already be observed in the early stages of frying-induced changes in the oil. A pre-production model is being developed at the time of writing (Figure 8.11). A general note on this successful instrument development is that chemical changes are often accompanied by physical changes that are much easier to measure and, in some cases, are even better indicators than the chemical variables that can be realistically measured in an industrial application.

8.8 Discussion

Significant progress has been achieved in the commercial development of chemical sensors in the last few years. For specific applications, laboratory prototypes have been turned into practical devices for medical or industrial use. It is, however, clear that the development of each novel commercial sensor is a costly and lengthy process. This is relevant particularly for the food industry where a small market only exists for a particular sensor configuration for most of the applications of interest. There is a wide variation in analytes, analyte concentrations, pH values, operating temperatures, sample viscosities, ambient humidities, interfering chemical compounds, suspended particles and many other variables in food applications.

Only chemical sensors for applications with a particularly favourable market position (wide applicability, economical or safety implication of the measurement) will be fully optimised. For a number of further applications, sensors of a related physical property can be developed instead. For many other applications, the best use of general purpose sensors will be made with the help of techniques such as robotic sampling facilities installed on a production line together with a sample conditioning appliance and the sensor in the bleed line (Figure 8.12). Flow injection analysis (FIA) techniques are being developed for use in such facilities.

Multi-variable measurements together with new techniques of evaluating such signals will also play an important role. This helps both in overcoming the problem of cross-sensitivity and also in gaining a fuller picture of complex process characteristics to be monitored. Techniques that are being developed for this purpose include neural networks and increasingly fuzzy logic. It must be kept in mind, however, that sophisticated data evaluation procedures are no replacement for a set of sensors with a sound physicochemical relationship with the properties of interest,

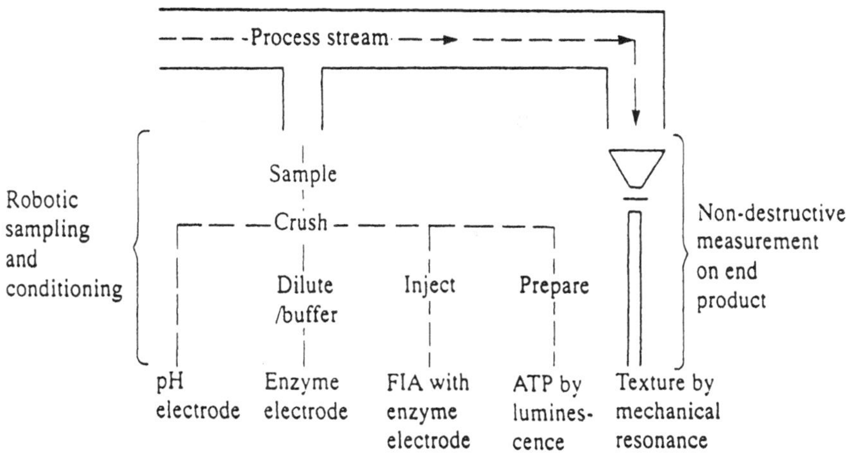

Figure 8.12 Example of the robotics approach. From Kress-Rogers (1993c).

but are complementary to them. Without such sensors, an efficient and robust calibration is not possible.

The scope of this chapter allows only for a brief introduction to chemical sensors. A more detailed description of the underlying principles and application details of chemical sensors is given in Kress-Rogers (1993a, b). Further references and a bibliography are also found there.

References

Arnold, M.A. (1990) Fiber-optic biosensors. *J. Biotechnol.*, **15**, 219–228.

Arnold, M.A. and Meyerhoff, M.E. (1988) Recent advances in the development and analytical applications of biosensing probes. *CRC Crit. Rev. Analyt. Chem.*, **20**, 149–196.

Badley R.A., Drake, R.A.L., Shanks, I.A., Smith, A.M. and Stephenson, P.R. (1987) Optical biosensor for immunoassays: the fluorescence capillary-fill device. *Phil. Trans. Roy. Soc. London*, **B316**, 143–160. (Presented by I.A. Shanks, 1986)

Ballantine, D.S. Jr and Wohltjen, H. (1989) Surface acoustic wave devices for chemical analysis, *Analyt. Chem.*, **61**, 704A–715A.

Bergveld, P. (1970) Development of an ion-sensitive solid-state device for neurophysiological measurements, *IEEE Trans. Biomed. Eng. BME*-**17**, 70–71.

Bergveld, P. (1988) Development and application of chemical sensors in liquids. In *Sensors and Sensory Systems for Advanced Robots*, NATO ASI Series, Vol. F43, (Ed. Dario P.) Springer Verlag, Berlin.

Bradley, J. and Schmid, R.D. (1991) Optimisation of a biosensor for *in situ* fermentation monitoring of glucose concentration. *Biosensors and Bioelectronics*, **6**, 669–674.

Brand, U., Reinhardt, B., Ruether, F., Scheper, T. and Schuegerl, K. (1991) Bio-field-effect transistors for process control in biotechnology. *Sensors and Actuators*, **B4**, 315–318.

Caras, S. and Janata, J. (1980) Field effect transistor sensitive to penicillin. *Analyt. Chem.*, **52**, 1935–1937.

Cass, A.E.G., Davis, G., Francis, G.D., Hill, H.A.O., Aston, W.J., Higgins, I.J., Plotkin, E.V., Scott, L.D.L. and Turner, A.P.F. (1984) Ferrocene-mediated enzyme electrode for amperometric determination of glucose. *Analyt. Chem.* **56**, 671–677.

Clarke, D.J., Blake-Coleman, B.C. and Calder, M.R. (1987) Principles and potential of piezoelectric transducers and acoustical techniques. In *Biosensors, Fundamentals and Applications* (eds Turner, A.P.F., Karube, I, and Wilson, G.S). Oxford University Press, New York, pp. 551–571.

Cleland, N. Hoernsten, E.G., Elwing, H., Enfors, S. and Lundstroem, I. (1984) Measurement of hydrogen evolution by oxygen-limited *Escherichia coli* by means of a hydrogen sensitive Pd-MOS sensor. *Appl. Microbiol. Biotechnol.* **20**, 268–270.

Daniels, P.B., Deacon, J.K., Eddowes, M.J. and Pedley, D.G. (1988) Surface plasmon resonance applied to immunosensing. *Sensors and Actuators*, **15**, 11–18.

Danielsson, B. (1990) Calorimetric biosensors. *J. Biotechnol.*, **15**, 187–200.

Danielsson, B. and Winquist, F. (1989) Biosensors based on semiconductor gas sensors. In *Biosensors: Fundamentals and Applications* (eds Turner, A.P.F., Karube, I. and Wilson, G.). Oxford University Press, pp. 531–548.

D'Amico, A. and Verona, E. (1989) SAW sensors. *Sensors and Actuators*, **17**, 55–66.

D'Costa, E., Dillon, M., Hodgson, F.J.A. and Quantick, P.C. (1988) Rapid prediction of banana fruit maturation using a glucose biosensor. *Analyst*, **113**, 225–228.

Den Reijer, M. (1990) Goede pH-meters verkriegbaar. *Vleesdistributie en vleestechnologie*, **25**, 34–39.

Eikelenboom, G. (1990a) Toepassingen van de pH-meting bij vers vlees. *Vleesdistrib. vleestechnol.*, **25**, 19–23.

Eikelenboom, G. (1990b) Het meten van de pH van vlees. *Vleesdistrib. vleestechnol.*, **25**, 24–33.

Gardner, J.W. (1991) Detection of vapours and odours from a multisensor array using pattern recognition; part I: principal component and cluster analysis. *Sensors and Actuators*, **B4**, 109–115.

Gill, C.O. (1976) Substrate limitation of bacterial growth at meat surfaces. *J. Appl. Bacteriol.*, **41**, 401–410.

Gotoh, M., Tamiya, E., Karube, I. and Kagawa, Y. (1986) A microsensor for adenosine-5'-triphosphate (using a) pH-sensitive field effect transistor. *Analyt. Chim. Acta*, **187**, 287–291.

Guilbault, G.G. and Jordan, J.M. (1988) Analytical uses of piezoelectric crystals: a review. *CRC Crit. Rev. Analyt. Chem.*, **19**, 1–28.

Karube, I. and Tamiya, E. (1989) Microbiosensors for clinical and process analysis. In *Bioproducts and Bioprocesses* (eds Fiechter, A., Okada, H. and Tanner, R.D.) Springer Verlag, Heidelberg, pp. 297–306.

Karube, I., Sode, K. and Tamiya, E. (1990) Microbiosensors. *J. Biotechnol.* **15**, 267–282.

Kimura, J. and Kuriyama, T. (1990) FET biosensors, *J. Biotechnol.*, **15**, 239–254.

Klein, M. (1991) Calcium-sensitive field effect transistor with inorganic layer. *Sensors and Actuators*, **B4**, 141–144.

Kohl, D. (1989) Catalytic reactions and electronic processes relevant in gas sensing: an extended abstract. *Sensors and Actuators*, **17**, 309–311.

Koudelka, M., Gernet, S. and De Rooij, N.F. (1989) Planar amperometric enzyme-based glucose microelectrode. *Sensors and Actuators*, **18**, 157–165.

Kress-Rogers, E. (1985) *Technology Transfer, Part II. The New Generation of Sensors.* LFRA Scientific and Technical Surveys, No. 150.

Kress-Rogers, E. (1986) Sensors for measurement of food properties and for quality control. *COST 91bis Sub-group Workshop*. 25 November 1986, Chipping Campden. (A summary can be found on pp. 11–14 of *Berichte der Bundesforschungsanstalt fuer Ernaehrung*, BFE-R-87-01 Sensors and Measurement of Product Properties—Instrumentation and Process Control. (ed. Paulus, K.O.) February 87.)

Kress-Rogers, E. (1987) Sensors and their applications to meat technology, In *Proc. Trends Modern Meat Technology II*, November 1987, (eds Krol, B., van Roon, P.S. and Houben J.H.). Den Holder, Netherlands (Pudoc, Wageningen, 1988), pp. 33–39.

Kress-Rogers, E. (1978/1988) Viscosity sensor for assessment of oil quality during use. *Symposium 'Frying'* (25 Feb 1988 at the LFRA), LFRA Symposium Proceedings No. 35, pp. 54–60.

Kress-Rogers, E. (1991) Solid-state pH sensors for food applications, *Trends in Food Science & Technology*. Volume 2, pp. 320–324 (December 1991). (Note typesetting errors: drift should read 0.05 pH units per 24 hours on p.323, biosensors should read sensors in Table 1.)

Kress-Rogers, E. (1993a) Chemical sensors, biosensors and immunosensors. In *Instrumentation and Sensors for the Food Industry*. (ed. Kress-Rogers, E.). Butterworth-Heinemann, pp. 581–689. Due to appear March 1993. ISBN 0 7506 11537.

Kress-Rogers, E. (1993b) The marker concept: frying oil monitor and meat freshness sensor. In *Instrumentation and Sensors for the Food Industry* (ed. Kress-Rogers, E.). Butterworth-Heinemann, pp. 523–580. Due to appear February 1993. ISBN 0 7506 11537.

Kress-Rogers, E. (1993c) Instrumentation for food quality assurance, In *Instrumentation and Sensors for the Food Industry* (ed. Kress-Rogers, E.). Butterworth-Heinemann, pp. 1–36. Due to appear February 1993. ISBN 0 7506 11537.

Kress-Rogers, E. and D'Costa, E.J. (1986) Biosensors for the food industry. *Analyt. Proc.* **23**, 149–151.

Kress-Rogers, E. and Turner, A.P.F. (1986) Immunosensors based on acoustic, optical and bioelectrochemical devices and techniques. *Leatherhead Food RA Technical Notes* No 49. (See also *Immunoassays for Veterinary and Food Analysis* (eds Morris, B.A., Clifford, M.N. and Jackman, R.) (1988), Elsevier Applied Science Publishers (Proc. Symp. Adv. Immuno Assays for Veterinary and Food Analysis, University of Surrey, July 1986.)

Kress-Rogers, E., Sollars, J.E., D'Costa, E.J., Wood, J.M. and Turner, A.P.F. (1988) Meat freshness using a biosensor array. *Proc. 34th Internat. Congr. Meat Sci. Technol.*, (Brisbane Aug/Sept 1988), pp. 508–510.

Kress-Rogers, E., Gillatt, P.N. and Rossell, J.B. (1990) Development and evaluation of a novel sensor for the *in situ* assessment of frying oil quality. *Food Control*, **1**, 163–178.

Kress-Rogers, E., (1987/1988) Viscosity sensor for assessment of oil quality during use. *Symposium 'Frying'* (25 Feb 1988 at the LFRA), LFRA Symposium Proceedings No. 35, pp. 54–60.

Kress-Rogers, E. D'Costa, E.J., Sollars, J.E., Gibbs, P.A. and Turner, A.P.F. (1992) *In situ* measurement of meat freshness with a biosensor array. In *Proc. The Nestlé Meeting on Biosensors, Opportunities for the Food Industry* (Lausanne, May 1992), pp. 41–46.

Leiner, M.J.P. and Wolfbeis, O.S. (1991) Fiber optic pH sensors. In *Fibre Optic Chemical Sensors and Biosensors*, Volume I, (ed. Wolfbeis, O.S.). CRC Press, Boston, Chapter 8, pp. 359–384.

Liedberg, B. Nylander, C. and Lundstrom, I. (1983) Surface plasmon resonance for gas detection and biosensing. *Sensors and Actuators*, **4**, 299–304.

MacDonnell, M.B. and Vadgama, P. (1989) Membranes: separation principles and sensing. *Selective Electrode Rev.*, **11**, 17–67.

Mascini, M. (ed.) (1989) Strategies for *in-vivo* sensing. *Abstr. 2nd Workshop Biomed. Eng. Action Europ. Community on Chemical Sensors for* in-vivo *Monitoring*, Florence, 12–15 November.

McCallum, J.J. (1989) Piezoelectric devices for mass and chemical measurements: an update. *Analyst*, **114**, 1173–1189.

Muehlbauer, M.J. Guilbeau, E.J. and Towe, B.C. (1990) Applications and stability of a thermoelectric enzyme sensor. *Sensors and Actuators*, **B2**, 223–232.

Muramatsu, H., Tamiya, E. and Karube, I. (1989) Detection of odorants using lipid coated piezoelectric crystal resonators. *Analyt. Chim. Acta*, **225**, 399–408.

Nylander, C., Liedberg, B. and Lind, T. (1982) Gas detection by means of surface plasmon resonance. *Sensors and Actuators*, **3**, 79–88.

Olthuis, W., Van der Schoot, B.H., Chavez, F. and Bergveld, P. (1989) A dipstick sensor for coulometric acid-base titrations. *Sensors and Actuators*, **17**, 279–283.

Persaud, K.C. and Dodd, G.H. (1982) Analysis of discrimination mechanisms in the mammalian olfactory system using a model nose. *Nature, London*, **299**, 352–355.

Persaud, K.C. and Pelosi, P. (1992) Sensor arrays using conducting polymers for an artificial nose. In *Electronic Noses, NATO ASI Series* (ed. Gardner, P.G., Dodd, G.H., Bartlett, J.) Kluwer Press (Proc. of NATO Workshop on Electronic Noses, Reykjavik, Iceland, August 1991), in press.

Sibbald, A., Whalley, P.D. and Covington, A.K. (1984) A miniature flow-through cell with a four-function CHEMFET integrated circuit for simultaneous measurements of potassium, hydrogen, calcium and sodium ions. *Analyt. Chim. Acta*, **159**, 47–62.

Sundgren, H., Lundstroem, I., Winquist, F., Lukkari, I., Carlsson, R. and Wold, S. (1990)

Evaluation of a multiple gas mixture with a simple MOSFET gas sensor array and pattern recognition. *Sensors and Actuators*, **B2**, 115–123.

Turner, A.P.F., Hendry, S.P. and Cardosi, M.F. (1987) Tetrathiafulvalene: a new mediator for amperometric biosensors. In *The World Biotechnology Report on Biosensors, Instrumentation and Processing I*. Online Publications, Pinner, UK, pp. 125–137.

Updike, S.J. and Hicks, G.P. (1967) The enzyme electrode. *Nature*, **214**, 986.

Vadgama, P. (1990) Biosensors: adaptation for practical use. *Sensors and Actuators*, **B1**, 1–7.

Van den Berg, A., Grisel, A., and Verney-Norberg, E. (1991) An ISFET-based calcium sensor using a photopolymerised polysiloxane membrane. *Sensors and Actuators*, **B4**, 235–238.

Van der Schoot, B.H. and Bergveld, P. (1985) An ISFET-based microlitre titrator: integration of a chemical sensor-actuator system. *Sensors and Actuators*, **8**, 11–22.

Walt, D.R., Barnard, S.M. and Luo, S. (1991) Optical immunosensors using controlled release polymers, *Symp. Am. Chem. Soc.* Division PMSE (Polymeric Materials Science and Engineering), Session on 'Biosensors: novel applications of polymeric materials,' Atlanta, April 1991.

Watanabe, E., Endo, H., Hayashi, T. and Toyama, K. (1986) Simultaneous determination of hypoxanthine and inosine with an enzyme sensor. *Biosensors*, **2**, 235–244.

Wolfbeis, O.S. (ed.) (1991) *Fiber Optic Chemical Sensors and Biosensors*, Volumes. I and II. CRC Press, Boston, London.

Index